THE WRONG STUFF

THE WRONG STUFF

HOW THE SOVIET SPACE PROGRAM CRASHED AND BURNED

John Strausbaugh

PUBLICAFFAIRS

New York

PublicAffairs
Hachette Book Group
1290 Avenue of the Americas, New York, NY 10104
www.publicaffairsbooks.com
@Public_Affairs

Printed in the United States of America

First Edition: June 2024

Published by PublicAffairs, an imprint of Hachette Book Group, Inc. The PublicAffairs
name and logo is a registered trademark of the Hachette Book Group.

The Hachette Speakers Bureau provides a wide range of authors for speaking events.
To find out more, go to hachettespeakersbureau.com or email
HachetteSpeakers@hbgusa.com.

PublicAffairs books may be purchased in bulk for business, educational,
or promotional use. For more information, please contact your local bookseller or the
Hachette Book Group Special Markets Department at special.markets@hbgusa.com.

The publisher is not responsible for websites (or their content) that are not owned
by the publisher.

Print book interior design by Bart Dawson

Library of Congress Cataloging-in-Publication Data
Names: Strausbaugh, John, author.
Title: The wrong stuff : how the Soviet space program crashed and
burned / John Strausbaugh.
Description: First edition. | New York : PublicAffairs, 2024. | Includes
bibliographical references.
Identifiers: LCCN 2023046976 | ISBN 9781541703346 (hardcover) |
ISBN 9781541703360 (ebook)
Subjects: LCSH: Astronautics—Soviet Union. | Astronautics and state—Soviet Union. |
Space race—Soviet Union. | Space race—United States—History—20th century.
Classification: LCC TL789.8.S65 S77 2024 | DDC 629.40947—dc23/eng/20240128
LC record available at https://lccn.loc.gov/2023046976

ISBNs: 9781541703346 (hardcover), 9781541703360 (ebook)

LSC-H

Printing 1, 2024

CONTENTS

1

THREE STOOGES IN SPACE

On the morning of October 12, 1964, a drab green bus pulled up near a launchpad at the Soviet spaceport called the Baikonur Cosmodrome in bleak and dreary Kazakhstan. The door opened and three small men in soft white aviator caps and what looked like wool leisure suits stepped down. They were dressed more for a cruise ship than a spaceship. They were very serious, almost grim, as they filed toward the launchpad, like men walking to the gallows, and with good reason. A monstrous rocket loomed ten stories high on the launchpad, steaming and hissing like a giant boiler ready to burst, towering above the rusty sands of the surrounding desert. Standing under it, another small man, this one in a spiffy Soviet air force uniform, hugged each of the three in turn, kissing their cheeks, acting far more cheerful than they were. He could afford to be. He was the cosmonaut Yuri Gagarin. He'd been the first man in space back in 1961. He'd survived his near-death experience in the same sort of space capsule they were about to fly in. He'd thrown himself into space and lived to tell about it. Now it was their turn to try their luck.

The trio rode up to the top of the rocket in an open elevator. One at a time, they wriggled through a small hatch in the rocket's pointy nose, then stuffed themselves like Vienna sausages into a ridiculously tiny space capsule. They were packed so tight they were practically in one another's laps, elbows tucked in, no room to move anything but their heads. It's hard to resist calling them, in the parlance of the time, three Commie stooges. Three small Soviet men in leisure suits waiting to be fired into space without helmets, without space suits, with a limited supply of oxygen, with virtually no safety gear, and with no way to exit the capsule quickly if the launch went wrong. Which, given the state of Soviet space technology in 1964, it certainly could. The model of rocket they were sitting on top of, called an R-7, was notoriously tetchy, with a record of spectacular and catastrophic launch explosions.

Why were they risking their lives in this way? Was there a scientific rationale for this flight? No. Would it be a step forward in the progress of manned spaceflight? Not really. This mission was entirely a political one. Since the launch of Sputnik in 1957, Soviet rockets had given Nikita Khrushchev, leader of the USSR, opportunity after opportunity to troll the Americans about their inferior and tardy space program. It had become a kind of drug to him, and he was addicted to it. The three men squeezed into the tip of the R-7 monster that morning were there because Khrushchev was jonesing for another chance to show up the Americans.

Six months earlier, Khrushchev had been upset by NASA's announcement that its new Gemini program would put a two-man capsule in orbit by late 1964 or early 1965. This would be a first—in space race terms, NASA's first first. The Americans had trailed the Soviet space program until then, and Khrushchev wanted it to stay that way. He summoned the man who'd given him the Sputnik and Gagarin victories, Chief Designer Sergei Pavlovich Korolev. (Though the standard English spelling is Khrushchev and Korolev, their names were pronounced like "KROOSH-chov" and "Ka-ROLL-yov." While we're at it, Russians pronounce Moscow neither like "Moss-cow" or "Mosk-owe" but more like "Moskva.")

They met in the Kremlin, in Khrushchev's comfy, unostentatious office. If the Americans were going to put two people in space by the end of the year, Khrushchev said, then he wanted to launch *three* people, and do it before the American launch—by the anniversary of the Bolshevik Revolution on November 7 would do nicely. (The revolution had occurred in October 1917 by the old Julian calendar, but when Russia adopted the Gregorian calendar in 1918 the date shifted into November.) Korolev tried to explain in technical detail that it would take much longer than that to develop a three-man vehicle and a rocket powerful enough to lift it into orbit. Khrushchev, characterized as a "semi-illiterate petty tyrant" by the science writer Leonid Vladimirov,[*] had grown up a farm boy and factory worker and confessed that his grasp of engineering was not good. He

---

[*] Vladimirov was the nom de plume of Leonid Finkelstein, a Gulag survivor who defected to London in 1966 and went on to a very long career in journalism there. His book *The Russian Space Bluff* was published in 1973 and poorly received because no one at the time believed the Soviet space program could be as slapdash as he described.

liked his chief designer, who had given him some great propaganda coups over the Americans. But he wanted another. So when Korolev told him that what he asked for couldn't be done, Khrushchev in effect replied, "Make it so."

Korolev did not argue further. Khrushchev may not have been a genocidal maniac like his predecessor Stalin, but he was not above shortening your life by sending you to Siberia or somewhere else unpleasant. Korolev had survived Siberia once already, during Stalin's mad reign of terror. Trudging glumly back to his top secret rocket design facility just outside Moscow, he explained Comrade Khrushchev's desires to his team of scientists and engineers. After unanimously declaring the task impossible, they all got down to work. None of them wanted to be sent to Siberia either.

There was indeed no time to develop a new three-person vehicle or a rocket big enough to launch it. They could only try to convert the one-man Vostok (East) capsule they'd been using since Gagarin's historic flight in 1961. The Vostok was like a large BB pellet, more space ordnance than spacecraft, an aluminum sphere roughly eight feet in diameter. When cosmonaut Alexei Leonov saw one for the first time, he was shocked that it was so small and *round*: "I had imagined a totally different type of space capsule, something more sleek and dynamic." He also couldn't get over how cramped it was inside. Like all cosmonauts, he was not tall, only five foot seven, but he didn't see how he would fit. In fact, Vostok was designed for a rider no taller than five foot six. And now they were planning to cram three of them into it.

Korolev's team stripped the Vostok interior nearly clean, then struggled to cram three seats in where there'd only been one. The seats were bucket recliners designed to help the passengers tolerate the punishing g-loads of takeoff. The engineers cocked them at angles, the middle one farther forward than the flanking ones, in a triangular formation. The seats just barely fit the space in this configuration, but there was obviously no way to stuff three cosmonauts in bulky spacesuits into a space that had already been cramped for just one. If they were sending Matryoshka nesting dolls into space they could stack them one cosmonaut inside another, but humans don't fit together so neatly.

One of Korolev's leading engineers, Konstantin Feoktistov, then suggested that three small men might fit *if they did not wear spacesuits*. Which meant that if the cabin depressurized, they'd die quick but agonizing deaths. When Korolev asked who'd be insane enough to do that, Feoktistov volunteered. He had stared death in the face before. He had been a teen when the Germans invaded in 1941. At sixteen he volunteered as a scout for the Red Army units defending his hometown. Caught, he was brought before a mass burial trench where a German officer was methodically executing prisoners with one Luger shot to the head each. Miraculously, when it came Feoktistov's turn the bullet only grazed his neck. He fell into the trench and played dead, then crawled out later when the coast was clear.

A fighter pilot and trained cosmonaut, air force Colonel Vladimir Komarov, was designated the pilot; like all cosmonauts, he was relatively short and light (as were American astronauts). Dr. Boris Yegorov, a physician on the team who monitored in-flight health telemetry, was the third crew member. Officially this was so that he could check the others' health during the flight, but really it was because he was also petite.

Even in their leisure suits the trio still couldn't be squeezed into the capsule, so Korolev asked for another sacrifice. The capsule's cosmonaut-ejecting apparatus had to go. They would be the first Soviet spacemen to land inside their capsule.

Why Yuri Gagarin and the other early cosmonauts did not land inside their Vostoks is partly a matter of geography, but more a matter of the neurotic Soviet obsession with secrecy. Project Mercury astronauts enjoyed the luxury of splashing down softly in the ocean, because the US was surrounded by temperate waters and had a very large navy for retrievals—thus Cape Canaveral. The Soviets put the Cosmodrome out in the middle of nowhere, where activities and especially failures were far from prying eyes. For such a vast land mass, the USSR had little access to waters that weren't icy, not ideal for landing a spacecraft in. The Cosmodrome was surrounded by vast tracts of flat, featureless, remote steppe, and that's where Vostok capsules came down. But they were hard landings that could kill any cosmonaut inside, and eventually one did. So Vostok missions relied on ejector seats. When atmospheric pressure sensors in the descending vehicle registered an altitude of seven kilometers, they automatically triggered explosive bolts that blew the hatch cover above the cosmonaut's head. Then more explosive bolts under the cosmonaut's butt fired the ejector seat through the opening. As it shot clear of the capsule, the seat's parachute canopy opened. At four kilometers, the seat and its parachutes fell away, the cosmonaut's smaller parachute opened, and he or she drifted to the ground, as though they'd jumped from a passing plane instead of bursting out of a plummeting spacecraft. Parachute practice was an essential component of cosmonaut training. Valentina Tereshkova, who became the first woman in space in 1963, was chosen largely because she was an experienced parachutist.

The next time you feel inclined to go on about what space cowboys America's early astronauts were, you should picture their Soviet counterparts, male and female, being fired out of their hurtling BBs at high altitude. Not because it was good science, but because their bosses in the government wanted them to be far away from public view. The story of the Soviet space program is a litany of such oddball make-do workarounds, most of which the Soviets brought on themselves.

There was absolutely no room for three ejector seats in the three-man Voskhod 1 (Sunrise 1), as the scooped-out Vostok was named to pretend that it was something new. Instead, Feoktistov and others designed a new braking system with two parachutes and small retro-rockets. They tested it using a Vostok capsule that had previously been flown by Gherman Titov, the second cosmonaut in space after Yuri Gagarin, and had been on display in a museum. According to Vladimirov, three monkeys sat in for the humans. Others said it was three humanoid dummies. In either case, they were out of luck, because the parachutes failed and the capsule smashed into the ground at such velocity it was no good anymore as a museum display. If there were monkeys on board the impact killed them, just as it would cosmonauts. While struggling to fix the problem, members of Korolev's team took to calling Voskhod 1 the "space-grave." Vasily Mishin, a gifted designer who had been with Korolev since the postwar years, condemned the project as "a circus act." Likely he was thinking of a clown car.

It was determined that even in their leisure suits the cosmonauts weighed too much. They were put on a strict diet.

So it was that on October 12, 1964, the three small, half-starved Soviet men wearing no protective gear and with no hope of emergency evacuation squashed themselves into an oversize BB pellet and were fired off into space. It was said that Korolev was so nervous that he visibly trembled throughout the rocket's liftoff and ascent and only calmed down once the capsule achieved orbit. Everything about the mission was wrong, and yet, as they often did at this stage in their space program's history, the Soviets somehow pulled it off. Feoktistov and Yegorov both reported the odd sensation that they were hanging upside down in the capsule, an effect of the weight-lessness. And with an environmental system built for one passen-ger, the temperature and humidity in the capsule quickly went tropical. But otherwise the mission was remarkably problem-free throughout its twenty-four-hour, sixteen-orbit flight.

On the third orbit the trio spoke to Khrushchev by radiotele-phone. He was in his study at his dacha in the exclusive Black Sea resort Pitsunda. Typically a Russian dacha is a modest country cot-tage where city folks go to relax and weed cabbage. Fred Coleman, an American journalist posted to Moscow, described Khrushchev's dacha as "the finest estate in the Soviet Union. . . . Pine forests and a ten-foot-high concrete wall sealed off the huge seafront property from prying eyes. . . . The main house, a two-story mansion, con-tained priceless oriental rugs, a Japanese garden on the roof and an elevator running up an outside wall." Khrushchev, a terrible swim-mer (Mao made fun of him), nevertheless liked to dog-paddle in the glass-enclosed pool with the retractable roof. "Telephones were fixed to trees along the garden paths" where he took long walks. Even in a Communist state, it's good to be the king.

Soviet journalists filled the room with their television cameras and mics. Khrushchev was ebullient, actually bouncing up and down when he heard Komarov's crackling voice. He congratulated the cosmonauts and promised them "a welcome stronger than the force of gravity" when they came to Moscow in a few days.

On October 13 the world's first space trio banged down to the ground in northern Kazakhstan. Miraculously, the new braking measures worked, and they prized themselves out of the capsule shaken up but unharmed. Korolev was as surprised as anyone. "Is it really true that it's all over? The crew has returned from space without a single scratch?" Not for the first or last time, he had pried open the jaws of Defeat, stuck his head in, reached all the way down Defeat's throat, and yanked out Victory.

The space race had begun two decades earlier, as World War II in Europe was ending. In the spring of 1945 Soviet troops were flooding into Germany from the east and the Allies from the west. In the middle stood the rocket scientist Wernher von Braun (like Brown, not Brawn), designer of the V-2, member of the Nazi party, officer in the SS, user of slave labor from the concentration camps, and future hero of NASA and Disney. At Hitler's order von Braun had begun using V-2s to rain destruction on London and other targets the previous autumn. Both the Americans and the Soviets recognized that he had advanced past their own rocket developers, and each hoped to grab him and as many of his rockets as they could before the other side got to him. Von Braun considered the vengeful Cossack hordes rushing at him from one side and the gum-chewing Americans ambling toward him from the other and sagely decided he'd much rather be captured by the Americans. He and his top team sought them out and surrendered to them in Austria—a thousand scientists and engineers, along with a cornucopia of blueprints and detailed notes, and the parts for one hundred V-2s. The Soviets were left to scrounge for whatever scraps of the V-2 and lower-level engineers the Americans left behind.

The race for space between the US and the USSR was a severely uneven match in far more ways. America came out of the war the world's only superpower, the world's only atomic power, a triumphant military, industrial and financial colossus. The Great Patriotic War, as it was known there, left the USSR a smoldering ruin. As many as *27 million* Soviet soldiers and civilians were killed in the fight against the invading Germans, compared to some 420,000 American casualties. American territory was almost entirely pristine and untouched by the war; the Soviet Union had been ravaged, with thousands of villages, towns, and cities flattened, factories destroyed, rail lines broken, and millions of acres of farms turned wasteland. Horrendous famine, plague, and widespread homelessness followed. The Americans' Marshall Plan helped Europe get through its devastation, but Soviet leader Joseph Stalin refused the aid. At war's end those Soviet citizens who had managed to survive the madman Hitler still had to survive the madman Stalin, who massacred and brutalized millions of them on his own hook.

Yet somehow the Soviets managed to build themselves up into the world's *other* superpower with astonishing speed. They had their own atomic bomb by 1949—built with knowledge stolen for them by spies in the Manhattan Project, but still—and their own H-bomb by 1953, and their own long-range bombers to drop them from—copies of American bombers, but still. In a decade they appeared to have raised themselves up from ashes and rubble to challenge the US. The imposing facade of the Iron Curtain hid a lot that was tumbledown and bargain-basement, but the world didn't see that at the time. What the world saw was the Soviets pulling even with the Americans—and then, in the space race, shooting ahead. Beginning in 1957 with Sputnik, the first man-made satellite in space, the Soviet space program beat the Americans to one milestone after another. The Americans were eventually the first to step on the Moon, but it was the Soviets who put the first man in space, the first woman in space, the first objects on the Moon and Venus, completed the first space walk, and more. For the first several years of the space race they were universally seen as being in the lead, and happily basked in the world's adulation.

The Americans were stunned every time. How were the poor, backward, war-ravaged Commies beating them to the punch over and over? They would have been even more astonished if they knew how ramshackle and low-budget the Soviet space program really was. Or how many disastrous failures the Soviets hid for every victory they trumpeted. The Soviets used secrecy and lies to bluff the world about their space program, their nuclear arsenal, and much more. All governments keep secrets and tell lies, but the Soviets were maniacal about it. Secrecy had been a Bolshevik survival tactic during their years as an illegal underground, down to their use of false names like "Lenin" and "Stalin." They carried it on to extremes when they got into power. Operating their space program in secrecy had obvious propaganda advantages. They could exaggerate successes, like Voskhod, and hide all but the most obvious failures, while NASA ran its program out in the open, with many public embarrassments. The secrecy extended to Korolev himself.

The Soviets worried that if the Americans learned his identity they'd try to assassinate him, so he was known in public only as the Chief Designer until the day after he died.

Which is not to say the Americans didn't have their own secrets. The ostensibly civilian NASA was in many ways a fig leaf for and distraction from military activities in space. For every public NASA launch there were at least two secret military ones that we know of now. In secrecy, the Soviets could be more direct about military involvement. All their launches were conducted by specially trained military staff, the Strategic Rocket Forces, working side by side with civilian scientists and engineers.

Something else the Soviets were happy to hide was how fantastically dysfunctional they were. The truth was that if anyone was good at making Communism work, it wasn't the Soviets. With them it was just more tsarist-style big man autocracy in Marxist-Leninist drag, with Lenin and Stalin and Khrushchev in basically the same tyrannical roles as Ivan and Peter and Nicholas. For all its citizens and subjects except for a minuscule elite, every day, in ways large and small, Soviet Communism was a dismal, punishing failure. All economic activity was centrally planned in Moscow, overseeing everything from rockets to rolling pins, steel mills to shoes. The immense, multilayered bureaucracy created to carry out these plans was, like most bureaucracies only more so, a breeding ground for corruption, inefficiency, and cynical laziness. Departments and agencies spread like blight in a field of potatoes, with often overlapping responsibilities, duplicated labor, and waste of limited resources. Instead of obedient comrades marching shoulder to shoulder into the future, the system was in fact a vipers' nest of bluster, big egos, and backstabbing. Government agencies, historian William Burrows writes, ran like "fiefdoms that were . . . hopelessly inefficient and ensnared in a political system that was built on flattery of superiors, treachery, low-key but ferocious competition, and paranoia." The bureaucrats in charge were "mid- and high-level prima donnas suspicious and wary as foxes who suspected virtually everyone and trusted almost no one."

Soviet industry was capable of the spectacular, as it demonstrated flooding battlegrounds with military hardware during the Great Patriotic War, but was more usually mired in the mediocre. Factory managers were given quotas, stressing quantity over quality. The manager of the steel mill who met his mark for girders kept his job, and his head, even if the girders were so inferior they just went into a rusting pile. Because all workers' wages were centrally programmed by the bureaucrats, they had no incentives to work well and every incentive to pilfer whatever they could to sell on the thriving and ubiquitous black market. Supply lines across the vastness of the USSR were a hopeless tangle; factories often sat idle waiting for raw materials, just as shops sat empty waiting for (inferior) finished goods. In their paranoia, the Soviets built up the largest standing army in the world. Feeding and equipping it was a huge drain on the national budget. The giant state-run farms to which Stalin had driven all the peasants in the 1930s were so unproductive that the Communists were often forced to buy food from the hated, well-fed capitalists. The whole economy depended on Siberian oil, the one sector that produced, until it didn't. (Years later, Senator John McCain would famously snipe, "Russia is a gas station masquerading as a country.")

Contributing to the low productivity and general disorder was a ruinous level of drunkenness. In *Vodka Politics*, political scientist

Mark Lawrence Schrad explains how drunkenness was the scourge of Russian society, from the medieval Muscovy princes through the tsars to the Communists. Ivan, whom they didn't call the Terrible for nothing, was the first tsar to monopolize the production and sales of alcohol under strict government control, beginning a tradition of state-sponsored alcoholism that carried on right through the Soviet years. Keeping the masses plastered on state-produced vodka both filled the government's coffers and helped keep the masses docile, befuddled, and too monstrously hungover most times even to think about revolting. Even the Bolsheviks' world-changing revolution of 1917 featured crowds of rioters throwing open the wine cellars and liquor cabinets of the rich and getting so drunk that the streets of Petrograd were impassable with the bodies of the passed-out masses.

This continued right through to the end of the USSR. In the mid-1980s, recorded rates of consumption were literally staggering. Millions of drunks were swept off the streets every year and dropped at official drying-out stations. Whenever the government tried to sober up the nation by reducing the availability of state-sold vodka, sugar disappeared from all the shops because people used it to make samogon (moonshine) at home.*

It wasn't just the masses getting plastered. Russian political history is as much *Animal House* as *House of Cards*. Peter the Great called his drinking buddies the "All Mad, All Drunken, All Jesting Assembly." Their drunken debauches were legendary. They did a rock-stars tour of Europe and England that left all their

---

* A reporter once quoted Vladimir Putin saying, "There are three ways to influence people: blackmail, vodka, and threatening to kill." The quote can't be verified because the reporter died in a mysterious plane crash days later.

accommodations ripped up, smashed, and "filthy in the extreme." During World War II, Stalin hosted enormous banquets for visiting allies. Even Churchill, who famously drank morning, noon, and night himself, was overwhelmed by the endless rounds of vodka toasts. Many a lesser man slid under the banquet table, to be carried out later. After the war, Stalin forced his top ministers to get falling-down drunk every night. "Eternally suspicious of plots to do him in," Schrad writes, "Stalin used alcohol to keep his inner circle off balance: make his closest comrades (or potential rivals) drink to excess in order to draw out their honest opinions and lay bare their true intentions. . . . Soviet high politics assumed the air of a college frat party with the devil. . . . Before attending to their assigned duties in the morning—or, more often, early afternoon—the Soviet leadership staggered outside to vomit or soil themselves before being borne home by their guards."

Korolev and the others in the space program competed with far-off NASA while having to contend with all this disarray at home. They didn't have anything like the resources the Americans could put into their space program. They jerry-rigged and improvised like mad to work around inferior equipment, parts that were never delivered, intramural squabbling among departments. Their technology and computers were hopelessly behind. Where the Americans concentrated all their civilian space activities in the one entity NASA, Soviet leaders acted almost whimsically in spreading their support out among jealous rivals. Korolev would often find himself having to fend off competition from other designers. Instead of long-term goals, they lurched forward to satisfy their political leaders' hunger for bragging rights. They leaped into the void, taking risks NASA never would have permitted, like Voskhod.

Writing about the jet pilots who became America's first astronauts, the journalist Tom Wolfe popularized the notion of the Right Stuff, the jet pilot's "ability to go up in a hurtling piece of machinery and put his hide on the line and then have the moxie, the reflexes, the experience, the coolness, to pull it back at the last yawning moment—and then go up again *the next day. . . .*"

It was implicit in this transaction that the "hurtling piece of machinery" the pilot/astronaut flew would be the very best that American money and know-how could produce—in short, the best in the world. The greatest country in the world, the intellectual and technological powerhouse of nations, had his back. He knew that if he didn't fail, the tech probably wouldn't fail him.

Cosmonauts had no such assurances about the machinery that hurtled them into space. If the astronauts displayed a courage born of confidence in themselves and their country, the cosmonauts' bravado was more reckless and fatalistic. Maybe the greatest paradox of the space race was the apparent reversal of roles. You might expect the Americans to be the space cowboys, flapping their Stetsons and whooping "Yippee-ki-yay!" as they rode to the stars on bucking broncos with rockets up their patoots. And the Soviets to be grim, interchangeable Commie robots marching lockstep into space at the command of their Fearless Leader. The reality, as reality usually is, was far messier. If anyone looked like robots it was the legions of faceless NASA bureaucrats and defense industry pencil-pushers in the US. NASA proceeded cautiously and methodically, building their program one-small-step-for-man at a time, learning from their mistakes, plodding toward their long-range goals. Astronauts came off in public as bland and interchangeable models of the American Everyman, the Man in the Gray Flannel Spacesuit.

Compared to them the cosmonauts were the real buckaroos. And maybe that's to be expected. As Soviet citizens, they had grown up poor in a poor land, used to making do in threadbare circumstances. They had survived the cataclysm of the Great Patriotic War, and the horrors of Stalin before and after that. They'd been left standing when millions and millions of fellow citizens' lives were sacrificed all around them. And they drank vodka like water. A certain cavalier attitude toward putting their own hides on the line might be expected.

If the American program set out to prove, with meticulous care, that it had the Right Stuff, the rash, slapdash, yet undeniably successful efforts of the Soviet program demonstrated the reverse. Call it the Wrong Stuff.

2

THE POWER COUPLE

To get anything done in the Soviet Union you wanted excellent survival instincts, a domineering personality, unwavering ambition, or political clout. Between them, Nikita Khrushchev and Sergei Korolev, one of the signal power couples of the Cold War USSR, had it all.

Nikita Sergeievich Khrushchev was five foot one and looked about that wide, with a head like a peeled potato. He was born in 1894 in a tiny, abjectly poor peasant village near Ukraine. He went to work with his father in the farm fields after the most rudimentary education. "After a year or two of school, I had learned to count up to thirty and my father decided that was enough of schooling," he would recall. "He said that all I needed was to be able to count money, and I would never have more than thirty rubles to count anyway." Later they moved to the Donbas region of eastern Ukraine, where his father labored as a coal miner and Nikita, at fourteen, started working in a factory, then with his father in the mines.

After discovering politics and joining the Communist Party in the 1920s, the young Khrushchev maneuvered his way up the bureaucratic trellis to the Communist Party's Central Committee, running the capital region of Moscow. He would famously denounce Stalin after his death in 1953, but in the 1930s he worshipped him and craved his approval. So the former farm boy and part-time Ukrainian said nothing as Stalin drove the peasants onto state-run collective farms and confiscated their grain to the last kernel, causing the Holodomor, the famine in which millions of Ukrainians starved to death in 1932 and '33. Then, in Stalin's mad purges of that decade, Khrushchev dutifully rounded up tens of thousands of "enemies of the people," including his own friends and colleagues, and handed them over to "the meat grinder," as he put it, bombastically declaring that "we must march across the corpses of the enemy toward the good of the people." Stalin posted him back to Ukraine in 1938, beginning his most despicable period yet. There was a strong separatist movement in Ukraine, not surprisingly, and Stalin, through Khrushchev, set out to smash it. By the time the German army poured into Ukraine in 1941, Khrushchev had imprisoned or deported some 750,000 Ukrainians, and at least 300,000 had died. It was so bad in Khrushchev's Ukraine that some Jews who had fled the Nazis in Poland *asked to be sent back to Poland.*

As a political commissar with the Red Army during the war, Khrushchev played a minor role in the defense of Stalingrad, about which he exaggerated proudly for the rest of his life. He was back in Moscow by 1949, a member of Stalin's inner circle of thugs with

Georgy Malenkov, Nikolai Bulganin, the loathsome Lavrenty Beria, and Vyacheslav Molotov. All members of the Presidium, Stalin's reorganized and renamed version of the policymaking Politburo, they toadied and capered around their withdrawn, melancholy, and deeply paranoid Cheerless Leader, while jockeying for position and backstabbing one another. Khrushchev was now Stalin's "pet," as some of the others grumbled. Although he had improved his education a bit, he was no intellectual and considered by Stalin and the others a rustic and clown. "Without doubt, a man of little culture," Molotov sniffed. This was how he survived the era. Intellectuals and city slickers made Stalin uneasy, and when Stalin was uneasy he turned homicidal. Also, Khrushchev had the good grace to be significantly shorter than the sorely height-conscious leader—five foot one to Stalin's five foot five or six. (It was hard to tell exactly because he wore lifts, as the American photographer Margaret Bourke-White noticed when she got an audience with him in 1941. After seeing giant Big Brother images of him everywhere, she was startled by how short and weak-chested he was. Harry Truman, who was not tall himself, called him a "little squirt.")

Through Stalin's last few years, Khrushchev and the others plotted and girded for the power struggle that would be inevitable when he died, which he did in 1953. He had a stroke. Earlier that year he'd had the nine most highly respected doctors in Moscow, six of them Jewish, arrested on conspiracy charges. The best treatment the remaining doctors could come up with was to apply leeches and leave him lying in his urine-soaked clothes. If ever a man was hoist with his own petard.

Khrushchev took charge of the funeral arrangements. State film crews fanned out across the USSR to document the colossal hippodrome of crocodile tears he orchestrated around the nation, 200 million citizens in public rituals of mourning for the monster who had brutalized and massacred them for thirty years. As Khrushchev distanced himself from Stalin's legacy over the next couple of years he thought better of doing anything with this footage. It sat in a vault until the filmmaker Sergei Loznitsa discovered and edited it into the fascinating documentary *State Funeral*, released in 2019.

At first Khrushchev vied with Beria, Molotov, and Malenkov to succeed Stalin. Beria, as head of the NKVD—the secret police, later renamed the KGB and then the FSB—had files on everyone, making him the most dangerous adversary. Khrushchev and Malenkov, backed by Marshal Georgy Zhukov and the Red Army, had Beria arrested—Khrushchev would joke that he "dropped a load in his pants"—sentenced for treason and other high crimes, and shot, after which they prudently destroyed those files. Khrushchev was named First Secretary of the Central Committee, ostensibly the top dog, though it would take him a few years to outmaneuver the others and emerge as *capo di tutti capi.*

All were in over their heads. These were not great statesmen, political geniuses, or selfless servants of the people. As tyrants will, Stalin had surrounded himself with people who were no threat to his ego or his supremacy. These were men of no morals, a goon squad, henchmen, like a gaggle of Mafia soldiers whose godfather has died and left them the keys to his dark empire. They didn't succeed Stalin so much as simply survive him. Most of Khrushchev's problems would be of the self-inflicted kind.

★ ★ ★

Before he could partner with Khrushchev, Sergei Pavlovich Korolev had to survive Stalin too. He was built like a cinder block, with dark hair greased straight back on his large, square head. His brown eyes stared with eagle intensity from his wide face. His solid appearance hid how broken his body really was. Younger people who worked for him noticed that he never really laughed. He couldn't—his jaw had been broken and reset itself so that he couldn't open his mouth wide enough to laugh. When he did open his mouth at all he revealed jarringly phony-looking dentures. Those around him also found it uncanny that to look to one side he turned his whole torso, not just his head. He couldn't turn his head. His neck was permanently ruined. His heart was damaged as well. At times of stress—and he was usually under stress—it thundered in his chest and threatened to give out entirely. And he had painful bleeding intestinal ulcers.

A man who knew he was dangling on a short string, he rushed his staff as much as himself, and his frustration often boiled over. He promised Khrushchev anything to get what he wanted from him. Then he'd push his staff night and day to meet nearly impossible goals with cruel deadlines. If they disappointed him, one engineer later recalled, "he reacted furiously, verbally ground us into dust, called us imbeciles, and promised to have us dismissed. . . . For Korolev, rage was an art form."

He was programmed by life to be this tough son of a bitch. He was born to an unhappy marriage near Kyiv in Ukraine in 1907. His father left and his mother farmed him out to his grandparents. In his adolescence he survived the chaos after the Bolshevik Revolution of 1917, when Ukraine struggled for independence and was fought over by the armies of nationalists, Bolsheviks, White Russians, Germany, and more. In his teens, like a lot of youth at the time, he became enraptured by aviation, maybe as an escape from the wretchedness of life on the ground. He also found inspiration in the science fiction of Jules Verne and H. G. Wells, as did all the pioneers and early promoters of rocketry—Robert Goddard in America, von Braun in Germany, Arthur C. Clarke in England. But the Soviets, always keen to distinguish themselves from the West, insisted that the real godfathers of their cosmonautics were a trio of homegrown eccentrics and outliers. Call them the oddfathers. Nikolai Kibalchich was a bomb-making revolutionary; Konstantin Tsiolkovsky a deaf, self-taught village schoolteacher; and Nikolai Fedorov a utopian who saw space travel as a way of conquering death itself.

Kibalchich looked like the quintessential bearded, bomb-throwing assassin of the nineteenth century, which he pretty much was. Born in Ukraine in 1853, he was arrested in 1875 for intending to distribute banned publications like *The Communist Manifesto*. After awaiting charges in prison for *three years*, he drew a sentence of two months. If he hadn't been an anti-tsarist revolutionary before, he sure was now. He joined a nihilist group and started making bombs for them to throw. One of his bombs killed Tsar Alexander II in 1881. Kibalchich was in his prison cell waiting to be hanged when he drew a very crude and simple sketch that's said to be the first manned spaceship ever conceptualized, a solid-fuel, controlled-explosion rocket with the engine in the back and a crew compartment up front, pretty much as they would be built decades later. It accompanied a note that said in part, "I am writing this project in prison, a few days before my death. I believe in the practicability of my idea and this faith supports me in my terrible situation. . . . If my idea . . . is recognized as implementable, I will be happy with the fact that I have done a huge favor to my native land

and to humanity." He was executed soon after. His prison papers lay unseen in government archives until after the revolution, when Bolshevik researchers found and published them.

Tsiolkovsky was also working out "certain aspects of the problem of going into space by means of a reaction-based device such as a rocket" in the 1880s. He was born a forest ranger's son in 1857. A childhood bout of scarlet fever left him almost totally deaf. There are photos of him as an adult using comically big and long metal ear horns; people would shout into the other end and he could sometimes, barely, make out a few words. He never finished high school, but was a voracious bookworm who practically lived in libraries and taught himself science and higher math, enough to get a job teaching in a small village schoolhouse out in the boondocks. Isolated and surrounded by silence as he was, maybe it makes sense that his thoughts turned toward space. He wrote about weightlessness in the vacuum of space with uncanny accuracy in the 1880s, said to be the first person to theorize about it. More mystically, he speculated that in the weightlessness of space humans would be truly free and happy for the first time. "Man will not remain on the earth forever. In his pursuit of light and space he will at first timidly probe beyond the atmosphere, then conquer all of circumsolar space."

By the early 1900s he had thought through the basic principles of spaceflight, predicting in extraordinary detail the rockets and vehicles that cosmonauts and astronauts would be riding sixty years later. He wrote about ways to cushion human passengers against the g-forces of liftoff, and how to spin space stations to create artificial gravity, as well as man-made satellites, jet-propelled aircraft, and atomic and solar power. The Wright brothers hadn't even flown yet. The tsar's censors were never comfortable with Tsi-olkovsky's ideas, but when the Bolsheviks took over in 1918 they touted his genius. They republished his writings, which became very popular with starry-eyed young Soviets who needed distracting from their actual circumstances on Earth. A portrait of him hung behind Korolev's desk.

Nikolai Fedorov was not scientifically inclined. He was a museum librarian. But his ideas about space travel inspired Korolev and other Soviet rocketeers. Born in 1829, Fedorov was a leading proponent of Cosmism. To Cosmists, the goal of humanity was to conquer death, not only by making people currently living immortal, but also by resurrecting every single human who ever died and making them immortal as well. This was going to make Earth an awfully crowded place, so Fedorov and the Cosmists embraced colonizing other planets in spaceships as an imperative. As odd as Fedorov's ideas might sound, Tsiolkovsky found inspiration in them, and through him, Korolev.

Can you imagine NASA with founding fathers like these? It's very difficult to find comparable visions among the nuts-and-bolts Americans. Jack Parsons, the cofounder of the Jet Propulsion Laboratory who was a member of Aleister Crowley's occult OTO, might qualify as one. But in general the Americans were out tinkering in the barn while the Soviets were dreaming of colonizing the universe. The very first report issued by RAND, the American military think tank, in 1946 was titled "Preliminary Design of an Experimental World-Circling Spaceship." It sounds like Jules Verne, more 1896 than 1946. One American writer, the popular Edward Everett Hale, did envision an orbital satellite in the 1870s. It was made of bricks.

In 1926, news that Goddard had successfully launched a liquid-fuel rocket electrified budding rocketeers everywhere. It didn't matter that his tiny rocket flew for less than three seconds and forty feet. Amateur rocketry clubs sprang up all over. Korolev joined one called by the acronym GIRD that worked out of an apartment building basement in Moscow, piecing together their first small rockets out of scraps. Once when they needed to solder some wiring, a member of the group sneaked a silver teaspoon out of his home and they melted it down. He had to sneak the spoon because Soviet citizens were supposed to have turned over all precious metals to the government. As the Soviet Union flexed its international might, you could be arrested for what amounted to criminal possession of fancy cutlery. Another young member of the group, Valentin Glushko, would go on to have a huge impact on Soviet rocketry as well as on Korolev's life.

GIRD's first successful rocket flew in 1933. Like Goddard's it was a small rocket, weighing thirty-eight pounds, and a short flight, to a height of thirteen hundred feet in eighteen seconds. But the young Korolev was already crowing, "Soviet rockets must conquer space!" As the 1930s progressed there was growing interest in the military applications of rocketry. In the US, the Jet Propulsion Laboratory worked on ballistic missiles. In Germany, von Braun and his team began to design what would become the V-2. In the USSR, the government folded GIRD into a new military bureau, the Reaction Propulsion Scientific Research Institute (RNII), to develop rockets in weaponry. Glushko began to design rocket engines, while Korolev focused on designing what we'd call cruise missiles.

Then, starting in 1937, came the Great Terror. Having risen to power in a tumult, Stalin moved violently and frantically to consolidate power, purging party stalwarts, military officers, professionals, Jews, intelligentsia—anyone he deemed a threat. Millions of Soviet citizens were swept up in the madness. Perhaps a million were executed right away, and more would die in hellish prisons and Gulag labor camps. In Stalin's dark mind the embarrassingly poor performance of Soviet aircraft in the Spanish Civil War marked aeronautics designers, including rocket designers, for purging. Marshal Mikhail Tukhachevsky, the man responsible for RNII, a Red Army hero, was arrested by the NKVD. Tukhachevsky was tortured, signed a "confession," and was executed by a pistol shot to the back of the neck. His whole family was arrested and several of them killed. Soon some forty-five thousand other officers and political commissars would be purged from the army, including sixty-eight of the eighty-five highest-ranking—on the eve of a war with Hitler that everyone told him was coming. The Earless Leader refused to listen.

The accusations and arrests at RNII flowed down from Tukhachevsky. Glushko was picked up in March 1938 and sentenced to eight years in prison. In return for reduced charges he fingered Korolev, who was arrested in his apartment that June and charged with being "a member of an anti-Soviet underground counter-revolutionary organization." For the rest of his life Korolev protested his innocence and complained that Glushko had "foully slandered" him.

Many years later, Korolev would sit up late into the night with two of his favorite young cosmonauts and shock them with tales of his arrest and imprisonment, things they had never heard before. He told them that after his arrest, the NKVD beat him for names of other "traitors and saboteurs." When he asked for water, a guard slammed a jug into his head. After he'd endured a month of torture, his eyes were so swollen he couldn't see the confession they shoved between his bruised hands to sign. He was sentenced to ten years hard labor and moved around to various prisons before being sent in the autumn of 1939 to the notorious Kolyma region in arctic Siberia, where the worst of Stalin's labor camps were. He traveled there with other prisoners in the hold of a cargo ship to the small port of Magadan, and from there in cattle cars 150 miles inland to the camp. He survived five months of subarctic cold—regularly as low as -50 degrees Celsius, or -58 degrees Fahrenheit—starvation rations, and slave labor, digging the frozen ground for gold, cutting trees, pushing wheelbarrows. Stout as he was, his body was ravaged. Along with injuries from his arrest and torture he had a large cut on his head that turned into an ugly scar, and half of his teeth loosened and fell out from scurvy.

Ironically, Hitler saved Korolev's life. When the German army ran rampant across Western Europe in 1939 it occurred to Stalin that perhaps he had acted a bit rashly when he slaughtered and imprisoned all his best military minds and weapons designers. Surviving scientists and engineers weren't freed, but they were transferred from the deadly prisons and labor camps to more humane "special prisons" where they were ordered to resume their work on aircraft, missiles, and such. They sarcastically called these facilities *sharashka*, slang for a shady operation or scam. Aircraft designer Andrei Tupolev, imprisoned in 1937, was assigned in 1939 to run Special Prison No. 4, an aviation design sharashka outside Moscow. The best talents in aircraft design were sent there.

In Siberia, Korolev received orders to report there as well, but no travel arrangements were made for him. His trip there sounds like Russian folklore. In prison rags, he set out on frozen feet for Magadan, 150 miles. On the snowy road he once found a whole loaf of bread, still warm, like it had dropped from heaven. Farther along, he collapsed from exhaustion on a dirt path. A kindly old man found him, rubbed his ruined gums with herbs, and propped him up in the pale sun. He saw another miracle—a butterfly, inspiring him to stay alive. He made it to Magadan, and from there to Tupolev's sharashka in the spring of 1940. He was thirty-six when he arrived. "He looked terrible," another prisoner would recall. "He was emaciated and exhausted." His body was in fact permanently damaged, destined to give out too soon, thanks to Comrade Stalin. After helping Tupolev develop the very successful Tu-2 bomber, Korolev was transferred to another sharashka, this one run by none other than Glushko. They worked on small rocket engines to give added thrust to propeller planes at takeoff. In 1944 he and Glushko were freed, though it would take Korolev more than a decade to win official exoneration for his "crimes."

In the spring of 1945 Korolev and Glushko were among the rocket experts Stalin sent to Germany to snatch anything and anyone useful the Americans had left behind. An engineer named Boris Chertok was sent as well. His four volumes of memoirs, translated into English and published as *Rockets and People* in the 2000s, are a trove of insider's stories. Korolev, Chertok, and Glushko spent two years in Germany gathering what they could, including some two thousand rocket engineers, mostly not top tier, whom the Red Army sent to the Soviet Union against their will. The rocket men would keep them at arm's length. They were resentful, believing they could have easily built their own version of the V-2 before von Braun did, if Stalin hadn't derailed their work by throwing them all in prison.

In April 1947, Korolev returned from Germany to Moscow, where he faced the man who had derailed him. Stalin wanted rockets that could deliver nuclear weapons (which he did not yet have) to US territory. He chuckled that an effective intercontinental ballistic missile force tipped with nukes would be "a straitjacket for that noisy shopkeeper Harry Truman." He ordered Korolev to start testing captured and reconstructed V-2s as a first step.

Korolev still wasn't officially exonerated of the crimes he'd supposedly committed. One imagines he spoke cautiously when he replied that he could build a better rocket than the V-2, a true Soviet rocket. No, Stalin replied. Just reconstruct the V-2. He distrusted his own scientists, which made sense, since he had liquidated many of the best. But he was also following Russian tradition. As far back as Peter the Great in the 1600s the Russians had borrowed, bought, or stolen technology from the more advanced West. When the Soviets were developing their own atomic bomb, Stalin insisted that his scientists precisely follow the stolen blueprints of the American ones. No innovations, no deviation. For a long-range bomber to deliver this bomb, he ordered Tupolev to recreate down to the last rivet a B-29 that had fallen into Russian hands at the end of the war. Tupolev glumly complied. The result was the Tu-4. Displaying the bleak sarcasm Soviet intellectuals often relied on, Tupolev called the Tu-4 "a locally built Boeing product." Years later, the supersonic Tu-144 would be an inferior copy of the Concorde, duly nicknamed the Concordski. The Soviets would even build a knockoff Space Shuttle.[*]

Korolev had survived Stalin's madness once. He wasn't going to risk his dreams of conquering space now. He put his head down and got to work on those V-2s. It was the start of a remarkable run.

---

[*] The same went for cars. The first car and truck assembly lines were built by Ford. The Zis and Zil limousines party leaders were driven around in were reverse-engineered from Packards and Cadillacs. The smaller Vaz was a knockoff of a Fiat, but cost the Soviet worker the equivalent of a Lamborghini, and you had to wait nine or ten years to get it. Russians told a joke when there were no KGB snitches around. Ronald Reagan enjoyed repeating it. A Soviet worker puts his rubles down for a car and is told to come back to pick it up in ten years. He says, "Morning or afternoon?" The guy behind the counter says, "What difference will that make?" The worker replies, "The plumber is coming in the morning."

3

A FLEA-BITTEN CANAVERAL

We gawked at what he had to show us, as if we were a
bunch of sheep seeing a new gate for the first time. . . .
We were like peasants in a marketplace, walking around the
rocket, touching it, tapping it to see if it was sturdy enough—
we did everything but lick it to see how it tasted.

—NIKITA KHRUSHCHEV

Korolev and his team, called NII-88 at first, later OKB-1 (Exper-
iment and Design Bureau 1), went to work in a top secret compound in the village of Podlipki, which was renamed Kaliningrad in
a crude attempt to confuse Western intelligence. There was already
a Kaliningrad, a German warm water port previously known as

Königsberg that the Soviets seized as war spoils and renamed. So now there were two Kaliningrads, one thousand kilometers apart. Vladimirov claimed that the Soviets took this "mania for secrecy" to such ludicrous extremes that "on all geographical maps published in the USSR not one Soviet city or town is shown in its correct position. . . . Every one is moved to one side or another. . . ." It makes you wonder why they printed maps at all, especially since so few Soviet citizens were allowed to have them. It was an insult to Western intelligence to think that they'd be fooled by all this, and they weren't, yet the Soviets persisted. To put off spies and snoops on the ground as well, the area around Podlipki, although it was right outside the city, was left to look like war-battered, unreconstructed farmland, with unmarked dirt roads and tattered peasants staring from the doors of crooked shacks as you drove past. You'd never guess anything high-tech was going on in such a wasteland. Only the initiated knew the circuitous route that led eventually to an iron gate in a blank brick wall; only those with proper papers passed through to the dilapidated prewar munitions factory compound inside. Chertok recalled that after two years admiring the Germans' clean and modern facilities they were appalled to be working in such a filthy, falling-down wreck. They shivered over their slide rules in the grimy, unheated spaces, and bunked in shacks built before the war, when indoor plumbing was a sign of bourgeois decadence. To bathe they stood, shivering some more, in long lines at a public bathhouse. The staff dining rooms were "murky, cavernous halls with bare walls," one of Korolev's engineers later recalled. He'd found it humiliating to stand in line "holding a tray and an aluminum spoon, only to choke over some cold macaroni. . . . It still haunts me like a nightmare." Gradually they would transform the place along the model of the bright, sleek German ones they'd envied.

Soon they were ready to test-launch their first reconstructed V-2s. For that they traveled to a new missile facility in the far southwestern corner of the country, on the left bank of the lower Volga River. The site was named for a drab nearby hamlet, Kapustin Yar—Cabbage Patch. Kapustin Yar was not far from Stalingrad, where the Soviets, with horrific losses, had turned back the German invaders in 1943. That made it a fit place for testing V-2s, but as a launch facility it was bare bones. Rockets were launched from rudimentary concrete pads on the flat ground. Observation trenches were dug at presumably safe distances from the launchpads. The rocketeers lived in railway cars and squatted over outdoor latrine trenches. Instead of a control bunker, as there would be at later sites, they hunkered inside a captured German Panzerwagen, an armored car, for mission control. For larger meetings they gathered in an abandoned bus. Missiles were assembled in a wooden shed. In black-and-white photos Korolev and his team huddle against the bleak, cold landscape in long coats and fur hats. They look less like rocket scientists than forced labor. Which, it being Stalin's USSR, they effectively were.

At this stage the space race was still an arms race. Once both sides had nuclear bombs, their militaries competed to develop von Braun's rudimentary, short-range V-2s into bigger, better intercontinental ballistic missiles, ICBMs, that could drop nuclear warheads on the enemy. The American army had shipped von Braun and his team first to White Sands, New Mexico, where they test-launched V-2s from 1946 to 1951. Moved then to a rocket facility in Huntsville, Alabama, which people took to calling Hunsville, they evolved the V-2 into the brawnier Redstone missile, which was shipped down the coast to be test-launched from a Florida site destined to become world famous, Cape Canaveral.

At Kapustin Yar, Korolev, Chertok, and the other rocket people worked with their military launch team, and some of the Germans were on hand for their experience. By 1948 they were launching their own version of the V-2, the R-1 (Raketa 1). They launched the first of the bigger, better R-2 model the following year. After Stalin died in 1953, Khrushchev and the other survivors of his inner circle asked Korolev to bring them up to speed on his rockets. He showed them an R-2, which Khrushchev later joked they almost licked.

That's an interesting image, given that Korolev's deputy Leonid Voskresensky actually *did* lick an R-2 once. No one on Korolev's team was closer to the flinty Chief Designer than Voskresensky, and no one embodied the program's reckless bravado more. He was the most colorful character on Korolev's team, a bon vivant who was considered a connoisseur of fine food and drink by his fellows, yet also an Evel Knievel with motorcycles and a general taker of risks. He was no looker, paunchy with hangdog eyes, but carried himself with an aristocratic swagger and was known for wearing a red beret, which would have been an eccentric and cheeky fashion statement in 1950s America, and in the Communist empire was as loud as an air horn. That may all have stemmed from his being an outsider. As the son of a priest at a time when Stalin was wiping out the Orthodox Church, he had been kept out of the Communist Party and was swept up in Stalin's purges. He was thirty and working as an electrician in Tupelov's Special Prison No. 4 when he met Korolev. Like Korolev, his time as one of Stalin's prisoners had permanent effects on his heart. At tense prelaunch moments they both took Validol, a pill that was taken for everything from heart murmurs to seasickness and, dissolving under the tongue, gave its users a distinctive menthol scent. He was the only member of Korolev's team who was not university trained—not "burdened with a higher education," as he liked to say—because the party had frozen him out. More a mechanic than an engineer, he had a pragmatic relationship to the machinery that the eggheads couldn't match. For that reason Korolev made him his deputy and launch director. He was the only one whom Korolev allowed to call him Sergei only, rather than the more polite Sergei Pavlovich. He was also never afraid to speak up when he disagreed with the Chief Designer, which Korolev appreciated—until he didn't.

To say Voskresensky was hands-on with the rockets would be a terrible understatement. Chertok relates the R-2-licking tale in his voluminous memoirs. The warhead container mounted on the R-2 missile contained a radioactive liquid. "When detonated at a high altitude this liquid was supposed to disperse, settling in the form of lethal radioactive rain," Chertok writes. During launch preparation for one R-2,

a turbid liquid trickled out of the top of the missile standing on the launch pad. Evidently, the chamber containing the lethal liquid had sprung a leak. The entire launch team hurried to get away from the missile. But what were they to do with it? Voskresensky, never at a loss in critical situations during a launch, ambled up to the missile. With the launch crew watching where it had taken refuge a hundred meters away, Voskresensky climbed up the erector to the height of the tail section so that everyone could see him. He gracefully stretched out his hand and ran his finger down the side of the missile through the liquid trickling down. Then, turning to the dumbfounded spectators, he stuck out his tongue and placed his "radioactive" finger on it. After climbing down, Voskresensky sauntered over and said, "Guys, let's get to work! It tastes like crap, but it's harmless." . . . Nevertheless, that evening at the hotel he availed himself of an additional shot of alcohol to "neutralize the substance and to allay the terror" that he had endured.

In *Red Moon Rising*, Matthew Brzezinski relates a similar tale. Korolev's rockets used a highly volatile cryogenic (supercooled) mixture of liquid oxygen and kerosene for fuel. The pipes that fed the supercooled liquid oxygen to the rockets' tanks frequently sprung leaks. The fuel was so touchy that once fueling had begun, turning off the supply to seal a leak could lead to a disastrous explosion. Voskresensky would walk over to the leak, wrap his jaunty red beret around it, unzip his trousers, take out his penis, and piss on the beret. "The minus 297-degree liquid oxygen would freeze the urine on contact, sealing the leak."

Chertok tells another tale from Kapustin Yar. In 1953 Marshal Mitrofan Nedelin came to the site to watch the test-firing of a new mid-range tactical missile. A poster boy for the Red Army, Nedelin was a national hero and had won a chestful of medals as an artillery commander during the Great Patriotic War. Though stiff-necked and by-the-book, he was also something of a visionary, one of the first Soviet military men to see that ballistic missiles, not bombers, would be the optimal way to deliver nukes to targets. He marched into Kapustin Yar this day with a full retinue of other high-ranking officers, their own medals clinking and clanking on their breasts. They stood in a clump near the launchpad to get a good view.

"It never occurred to anyone that the missile might fly not only forward along the route toward the target but also in the opposite direction," Chertok writes.

For that reason, the trenches were unoccupied. Everyone preferred to enjoy the sunny day on the lush steppe not yet scorched by the sun. . . . The missile lifted off at precisely the appointed time. Thrusting out a reddish cloud and resting on a glaring, fiery plume, it lunged vertically upward. But about four seconds later it changed its mind, pulled a sort of airplane barrel-roll maneuver, and went into a dive headed, it would seem, straight for our fearless retinue. Standing straight up Nedelin loudly shouted, "Lie down!" All around him everyone dropped to the ground. I considered it humiliating to lie down before such a small missile (it was just five metric tons), and leapt behind the shed. I took cover just in time. An explosion rang out. Clods of earth knocked against the shed and the vehicles. That's when I really got scared. What had happened to those who hadn't taken cover?

As the cloud from the impact rolled away, there was Nedelin, standing dusty but straight, the others picking themselves up and brushing themselves off around him, amazed and chagrined. Nedelin would repeat this performance, with a very different outcome, a few years later.

★ ★ ★

In 1955, the Soviets began searching for a place to build a new launch facility for intercontinental missiles. Kapustin Yar was fine for short-range V-2s and mid-range R-2s but too small for launching much bigger ICBMs, and anyway it was too close to listening posts in Turkey, a NATO member. They wanted a remote location for secrecy, and for unobstructed radio guidance. Radio guidance was needed because reliable onboard inertial guidance systems eluded Soviet engineering at this point. A Soviet missile could only fly true and hit its target using signals sent in from ground stations placed about every five hundred miles along its route. This was obviously a huge flaw that needed to be corrected if they were ever going to target, say, the continental United States. Soviet spies were working tirelessly to steal the knowledge and technology. In the meantime the rocketeers needed a broad, flat firing range across which they could string these signal stations. Also, it would be best if there were minimal down-range populations for missiles and debris to drop on. Not that the government worried itself *too* much about this. Over the coming years the Soviets would poison many of their own citizens and vast areas of their own land and waters with carcinogenic and radioactive materials from rockets, not to mention nuclear bomb tests, biological weapons labs, and so on. The catastrophe at the shoddily built Chernobyl nuclear plant was just one instance on an appallingly long list.

The military took the lead in the site search, but Korolev told his people to join in unless they wanted to "end up beyond the Arctic Circle." The place they settled on was hardly better than that: a howlingly forlorn spot called Tyuratam in the Kyzylkum Desert of southern Kazakhstan, near the muddy, sluggishly meandering Syr Darya River. Tyuratam—Kazakh for "Tore's tomb," Tore said to be a local nobleman descended from Genghis Khan—was suitably distant from any population centers, had an old rail spur on the line from Moscow to distant Tashkent, and was surrounded by a vast, bleak wasteland. The train from Moscow, a 1,500-mile trip, took three days. Chertok described Tyuratam as "a small brick building proudly bearing the sign *Vokzal*, a modest water tower for steam engines, a dozen or so trees barely clinging to life, several cottages for the station personnel, and five mud huts with local Kazakhs living on who knows what. Endless desert was all around. . . . Our first impression was one of sorrow and melancholy from the sight of the dilapidated mud huts and dirty back streets. . . ."*

Besides a few railway workers, the inhabitants included a handful of geologists and prospectors searching, without success, for oil. Sometimes nomadic Kazakhs would approach in wavering mirages across the rust-colored sands and pass by the town herding shaggy ponies or ill-tempered camels. (In the 1930s, a supremely ignorant Stalin had sought to collectivize these nomads, the way he was, fatally, collectivizing tens of millions of farm peasants. By the time he gave up, almost half of them were dead, along with most of their livestock.) The weather was dreadful. In the winter temperatures dropped to -50 degrees Fahrenheit and snow drifted chest-high, except when it was being whipped straight into your face by gale-force winds. Spring brought some flowering life and color, but it was cruelly brief, followed by a blistering hot desert summer with dust storms that sandblasted your face and rasped in your

---

* Vokzal had been a generic name for Russian train stations since the 1800s. A British entrepreneur established an imitation of London's Vauxhall Pleasure Garden in St. Petersburg, with its own rail station, Vokzal, in the 1780s. The name spread down the lines as far as Tyuratam and beyond.

lungs. On top of that, the place was overrun by rats and fleas, presenting the risk of cholera and typhus, and by poisonous snakes and angry scorpions.

The Soviets were in a hurry to get the facility built and operating, and they threw gangs of workers at it. In Stalin's day, Chertok noted, they probably would have been slave laborers from the Gulag, but the Gulag system had largely been emptied out as part of Khrushchev's anti-Stalin reforms. Instead, five thousand soldiers were shipped in to construct the site, working around the clock and in all extremities of weather. They were mostly veterans of the Great Patriotic War, so at least they knew something about surviving infernal conditions.

Officially it was given the deceptive and bland name Scientific Test Range Site No. 5. Showing that they'd read and absorbed their Tsiolkovsky, the rocketeers preferred to call it the Cosmodrome, like a gateway to the universe. By either title everything about the place was a misery—except for the very top brass, including Korolev, who, as everywhere in the Soviet Union, enjoyed nicer accommodations than the proles. At first the soldiers slept on the ground, then in tents and wooden barracks. Korolev's engineers bunked at first in railway cars, with outdoor privies and no showers. Near the launchpad four small cabins were built, with spindly trees planted around them. They were the nicest digs at the Cosmodrome. One became famous as the place where Korolev slept—when he slept at all—while at the Cosmodrome. Another became as close to a sacred place of worship as the Communists would allow, because it was where Yuri Gagarin slept the night before his history-shattering flight.

Meanwhile, a small, Soviet-glum town slowly grew around the rail station. It was just called Site 10 at first, then the marginally homier Leninsk. It was a "closed town," a Soviet convention (continued in post-Soviet Russia) where only authorized personnel resided, generally because they were engaged in secret military or scientific work. It was ringed with barbed wire and armed guards, like a forced labor camp, which in many ways it was. Outsiders had to be highly vetted even to be told of its existence, let alone allowed in for a visit, and Leninsk residents who were allowed out were forbidden even to say its name. By the early 1960s it housed about 4,000 workers, military personnel and their families. At its peak in the 1980s it would be up to 100,000, with rows of grim apartment blocks, schools, parks, movie theaters, and a beach on the muddy river. The Cosmodrome was a bouncy 15-mile drive across the desert.

Not a soul enjoyed being there. Cyclones of sand blew up especially in spring and fall. "[S]and was everywhere in the air, on the roofs, in hallways and apartments, on your teeth, in your hair, nose and ears," a resident wrote. "On the balconies real sand dunes would form, like somebody was shoveling sand in there." It lay like a film on your soup and made your every blink gritty. Then there were the rodents who swarmed the buildings for a bit of warmth in the vicious winters. Soldiers shoveled buckets of powdered poison down ratholes. Breathing the poison dust made them as sick as the rats. Because clean drinking water was hard to come by, dysentery ran riot through the troops. To try to grab a little sleep on summer nights when the temperature regularly hovered above 100 degrees Fahrenheit, people wrapped themselves in wet sheets, but the moisture attracted swarms of scorpions and tarantulas from the parched desert. In the early days betting on scorpions fighting in emptied vodka bottles was considered high entertainment.

Vladimir Suvorov, grandson of a Red Army general and son of a high-ranking government minister who got swept up in Stalin's purges, was a documentary cinematographer assigned to cover the space program. He kept a secret diary, which could have gotten him into a great deal of trouble had it ever been discovered. The diary was the basis for a book, published in English in 1997 as *The First Manned Spaceflight*. When he was assigned to Tyuratam he asked cosmonaut Vladislav Volkov why this awful place had been chosen as the perfect site for the USSR's major space port. Volkov grimly joked that it was because it offered "the maximal coincidence of inconveniences." (We'll meet Volkov again under other trying circumstances.) Conditions "were enough to test the endurance of any human being," agreed cosmonaut Alexei Leonov. He saw one technician "being bitten on the neck by a spider. He collapsed and died within minutes."

NASA chose Cape Canaveral because it was in one of the least populated and touristed areas of Florida's Atlantic coast, Redneck Riviera, plagued by clouds of mosquitoes, fire ants, and no-see-ums, with a lousy beach hard as concrete good only for racing hot rods. Because it was so remote it was already home to the missile test range at Patrick Air Force Base (now Patrick Space Force Base). But Canaveral was a four-star resort compared to Tyuratam. If Canaveral was godforsaken, Tyuratam was forsaken even by the Devil and all his minions. Between the loneliness, the hideous conditions, and the pressure to get the Cosmodrome built fast, mental and physical breakdowns were regular occurrences.

To make the worst place on Earth even worse, alcohol was officially banned. Had the authorities also forbidden smoking, they'd have another revolution on their hands. Suvorov and Chertok both report that the good Communist workers of the Cosmodrome put a great deal of effort and ingenuity into ignoring and evading this rule at every turn. "The cult of drinking hard liquor," one engineer recalled, "began at Kapustin Yar and followed us wherever we went." Suvorov relates once filling out a requisition form for alcohol, ostensibly to clean the sand out of his crew's camera equipment. He took it to the "most easy-going" of the bosses, Voskresensky. Voskresensky studied the form. With a straight face he asked how many men were on the film crew and how big they were, then doubled the requested amount of hooch. Suvorov was a huge Voskresensky fan from that moment on. Voskresensky himself always managed to have a bottle or two of vodka or cognac on hand. Chertok recorded many occasions of the two of them getting drunk together. There was nothing else to do at Tyuratam but work and watch scorpions die.

There was something else about the place, something more ephemeral, that Chertok says "tormented" the rocket men in the beginning. Before they knew anything about Sputnik or a manned space program, Tyuratam was to them a place of death. They were creating a site for launching nuclear weapons on ICBMs. "If, God forbid, we are the first to fire from this site at the Americans," Voskresensky once said gloomily, "then there won't be a second launch." The Americans would retaliate, and they'd be obliterated.

Building a launchpad for those ICBMs was the first priority; building one in the middle of a desert was a Cyclopean task. To begin with, they dug a yawning pit and walled it with meters-deep concrete to channel the rockets' flames at takeoff. It was said at the time to be the deepest man-made hole in the world. Suvorov described it as "a deep hollow like a crater of a gigantic volcano." Three soldiers died during its construction when a dump truck went out of control and rolled down the steep concrete bowl to crush them. Over this they built their first concrete launchpad. Just three hundred feet away they sunk an underground launch control bunker, entered after descending a flight of sixty steps from the surface. One large room was filled with banks of monitors and other electronics, off which there was a smaller second room. There were four submarine-style periscopes, two in each room. Voskresensky, as Korolev's launch director, would watch the liftoffs through one in the main room. Suvorov says that Korolev couldn't bear to watch himself. He'd sit fretting in the side room, and come barging out at the hint of any problem. Voskresensky joked grimly that if the Americans ever did attack the site, at least he'd have a great view of the nuke that killed them. There was no smoking inside the bunker, which must have driven the chain-smokers like Voskresensky crazy. They had to go out and up those sixty steps to smoke outside.

About a mile and a half away they erected a lofty assembly shed. The rockets and vehicles were built at Korolev's "Kaliningrad" facility near Moscow and elsewhere, then brought to Tyuratam in parts by rail. Inside the assembly shed the rockets were put together lying on their sides on railcars. Then they were slowly rolled out to the launchpad, which took about an hour. Korolev started a tradition of pacing alongside his beloved creations with some of his team as they inched toward the pad. In Suvorov's film footage it looks like a space age funeral procession.

Gradually the Cosmodrome grew, and grew some more. Many more launchpads, spread out well away from one another so that a disastrous explosion at one wouldn't damage the others; more housing and assembly sheds; an airstrip; miles and miles of concrete roads and rail lines and high-power electrical lines connecting them all across three thousand square miles of treeless desert, an area roughly the size of and precisely as uninviting as Death Valley. The hell of it was that within five years of the start of the Cosmodrome the Soviets finally had onboard guidance systems for their rockets that made the string of radio relays across the desert unnecessary. You could build a cosmodrome just outside Moscow if you wanted. But by then it was too late. Too many hours and rubles had gone into building the place to abandon it just because everyone, literally everyone, who worked there hated it.

Other cosmodromes would be added later. In 1966, some high school students in the industrial town of Kettering in England, tracking satellites on a war surplus radio receiver as a school project, noted a new one traveling a polar orbit. Their announcement of the discovery was the first public hint in the West of another Soviet launch facility at a place called Plesetsk, far in the north—not quite beyond the Arctic Circle, but close—near the White Sea port of Archangel. Its location was suitable for launching ICBMs at the US over the North Pole, and also for putting surveillance satellites in polar orbit, like the one the students discovered. The Soviets finally acknowledged the existence of the Plesetsk Cosmodrome in the 1980s.

On the night of October 4, 1957, a rocket blasted off from Tyuratam in a roar of blazing flames. Although it was launched in secrecy, the whole world would soon know about it.

4

SPUTNIK, FLOPNIK, AND YOUR LITTLE DOG, TOO

On the night of Friday, October 4, 1957, Senate Majority Leader Lyndon B. Johnson was on his sprawling ranch on the Pedernales River, a ninety-minute drive west of Austin, Texas, where his favorite pastime was "deer-hunting." On LBJ's ranch that meant one of two activities. In the daytime, deer-hunting meant bouncing around the ranch in his white Lincoln Continental convertible, banging away with his shotgun at any hapless deer who strayed into view. At night it meant drinking with his rich, powerful friends in a glass-walled, air-conditioned hunting blind at the top of a forty-foot tower. Black waiters in formal attire and white gloves glided around the blind's carpeted dining room serving dinner and refilling glasses. After dinner you went out onto a shooting deck with your shotgun and your topped-up whiskey. Johnson would abruptly switch on blinding floodlights. Below, deer who had been nibbling oats put out to attract them froze and you blasted away. John F. Kennedy would later say that the only way you could call that sporting would be if the deer had shotguns too.

On the night of October 4, Johnson was too preoccupied to enjoy massacring deer. Instead of looking down at shining, unblinking eye bulbs, he was obsessed with what was passing over his head every ninety-six minutes: Sputnik, the first man-made satellite in history, put up there by the Soviets. He reputedly growled something like, "Damn it, I will not sleep under a Red Moon!"

How Khrushchev would have loved to hear that. On the night of October 4 he was attending a banquet held for him by Ukrainian regional officials in the ornate Mariinsky Palace in Kyiv. His son Sergei remembered a lot of boring chitchat about crop yields, a major preoccupation in Ukraine, the "bread basket of the Soviet Union." (A PhD with a budding career as a rocket engineer, Sergei, then twenty-two, was his father's pride and frequent companion. He later became his most dedicated biographer.) At around eleven o'clock an aide whispered in Khrushchev's ear and he left the dining hall. He returned, beaming. "I can tell you some very pleasant and important news," Sergei recalled him announcing. "Korolev just called." The Ukrainians regarded him blankly. "He's one of our missile designers," Khrushchev explained. Then, foolishly, maybe a bit tipsy: "Remember not to mention his name—it's classified." He announced that the Soviet Union had just launched the world's first artificial satellite. Sergei writes: "'The Americans have proclaimed to the whole world that they are getting ready to launch a satellite of the Earth. Theirs is only the size of an orange,' Father continued excitedly. 'We, on the other hand, have kept quiet, but now we have a satellite circling the planet. And not a little one, but one weighing eighty kilos.'" The Ukrainians, Sergei recalled, gazed at their leader dutifully, "but their faces revealed their indifference." They did not know from satellites, he said. If Khrushchev had talked about farm equipment they might have gotten excited, but talk of missiles and satellites was over their heads. Well, maybe. But there were a number of Ukrainian-borns in the space program, starting with Korolev himself. Maybe these Ukrainian officials had not forgotten how Nikita Khrushchev had once tormented their land and weren't about to act too impressed with anything he bragged to them about.

American President Dwight Eisenhower's public response was nonchalant. He called Sputnik "one small ball in the air. Something that does not raise my apprehensions, not one iota."

The American people reacted much the same way, at least at first. They found Sputnik mildly interesting but nothing to get too worked up about. They sat around their black-and-white TVs on the night of October 4 watching the debut of a sitcom soon to be an icon of 1950s American complacency, *Leave It to Beaver*. And *The Adventures of Jim Bowie*, and Lee Marvin in the cop show *M Squad*, and the sturdy family favorite *Life of Riley*.

The rest of the world was standing outside, watching the Soviets take the lead in the space race.

Sputnik started as a "doorknob conversation." It was a psychological trick salespeople used. You wait until the apparent end of a meeting, your hand on the doorknob, then "remember" to make one last pitch. The client, caught off guard, says yes reflexively, just to get you out that door. Korolev did it to Khrushchev to get Sputnik approved. The setting was a meeting at Korolev's "Kaliningrad" facility in February 1956. Khrushchev was still struggling to climb above the rest of the thugs and gangsters who'd survived Stalin, some of whom accompanied him this day, to become the top boss. He was afraid the Americans would blow it all up before he got the chance. He was particularly frightened of the half-mad Air Force General Curtis "Bombs Away" LeMay, model for Brigadier General Jack D. Ripper in *Dr. Strangelove*. LeMay had directed the firebombing of Tokyo and the atomic bombing of Hiroshima and Nagasaki in 1945. Now he enjoyed taunting the Soviets with the vast size and global reach of the mighty Strategic Air Command nuclear bomber force he commanded—B-47s at first, later B-52s, neither of which the Soviets could hope to match. LeMay periodically sent fully loaded thermonuclear sorties into Soviet air space just to prove how puny and porous Russian defenses were, and would later sigh that he regretted not starting World War III and wiping the Commies off the face of the Earth when he had the chance. "Pax Atomica" he called it. Since he first saw, and resisted the urge to lick, one of Korolev's R-2s three years earlier, Khrushchev had become convinced that nuclear warheads on ICBMs would be the only salvation of his country. He came to Korolev's shop that day hoping that the Chief Designer had the goods.

Korolev knew how to put on a show for his bosses. ("In the Soviet Union, where everything depends on the word of some highly placed ignoramus, it was the only path to success," Vladimirov wrote.) In his Sputnik book *Red Moon Rising*, Matthew Brzezinski describes how, in OKB-1's giant white hangar space, Korolev showed his guests three rockets lying on their sides. It was a class in the development of Soviet rocketry, with visual aids. The first was an R-1, the Russian V-2, old hat to them by then. Next to it lay an R-2, the larger and souped-up descendant of the V-2 that had been the best Korolev had to show them back in 1953. So also old and familiar. But the third was something new: the longer, pencil-shaped R-5. It was twice as fast as an R-1 and far more powerful, and even more importantly, it was the first Soviet rocket built to carry a nuclear warhead. That very month an R-5 propelled a warhead 750 miles southeast from Kapustin Yar to a target site in the vast, empty Karakum Desert in Turkmenistan. The explosion was the equivalent of six Hiroshimas. Although not as widely touted as Sputnik would be later, it was historic in its own right: the first time a nuke carried by a ballistic missile was detonated. The Americans wouldn't manage that for more than a year.

Khrushchev had the military show off an R-5 in that year's May Day parade in Moscow. But much as Khrushchev loved it, he understood immediately that the R-5 was an intermediate-range missile and no threat to the Americans unless he could park it somewhere close (which he would do a few years later, triggering the Cuban Missile Crisis). Where were the true ICBMs to rattle at the Yankees? Korolev said the equivalent of "but wait, there's more!" With a theatrical flourish the hangar doors were rolled open, Korolev and his visitors stepped through, and his visitors' jaws dropped. An entirely new type of rocket towered above them. Not lying down, but standing up. And up. Even Sergei Khrushchev, the budding rocket scientist, was flabbergasted. Officially it was designated the R-7. Korolev's team affectionately called it *Semyorka*, more or less "Ol' Number Seven." It was a monster, tall as a ten-story building. While the Americans were developing their pencil-shaped rockets with one large engine, the R-7 had one central engine and four boosters strapped on around it like a billowing skirt. This configuration helped hurl the R-7 into the sky, then at about a minute and a half the four strap-ons fell away and the central engine served as a second stage, lofting it the rest of the way. Korolev told his visitors that the R-7 could carry a 3,000-kilogram warhead as far as 5,500 miles. Which meant it could hit targets in America.

Khrushchev was delighted and relieved. He had his ICBM.

Except he didn't. Korolev was in fact selling Khrushchev a Potemkin warhead. The R-7 was nowhere near ready to fly in February 1956, and when it did fly it was never going to be of much use as an ICBM. ICBMs were supposed to be ready to launch at a moment's notice. Soon enough the Americans would have ICBMs that could reach Soviet targets in *thirty minutes*. The R-7 was a very complicated machine, twitchy and testy as a thoroughbred. "Glitches occurred on an hourly basis," Chertok wrote. Besides how complex and tetchy the rocket was, there was the problem of the fuel. Korolev insisted on using liquid oxygen and kerosene, partly because Glushko argued against it. Because cryogenic fuels tended to be highly volatile, rockets were fueled slowly and with great care, and it was unsafe to leave liquid-fueled rockets filled and standing around. This made the R-7 effectively useless in case of attack. Solid fuel rockets could stand around ready to fire for a year or more, an obvious advantage in a nuclear war. As a nuclear weapons delivery system, the big, slow, unreliable Semyorka was not much good.

Finally, the R-7 was hugely expensive. Later, when Khrushchev heard the cost, he would cry, "We will be without pants!"

But for now he was sold, and thoroughly prepped for Korolev's doorknob conversation. Having made Khrushchev all giddy, Korolev casually mentioned that in addition to killing capitalists the R-7 could be used to hoist a satellite into orbit. Khrushchev's eyes said, "Yes, and?" And the Americans are working on their own, Korolev said. But we can beat them to it. Think of the propaganda victory! The first satellite in history, built by Soviet workers' hands! Khrushchev liked the sound of that. He asked Korolev if it would distract him from building ICBMs. Not at all, Korolev replied. It'll be the same rocket, just with a different payload.

Khrushchev thought about it a second and shrugged. Okay, why not? Make it so.

It was Korolev's turn to feel giddy. For the next year he drove his team relentlessly to fulfill the promise. The first R-7's components were hauled out to Tyuratam by rail in the spring of 1957. Korolev's team assembled it there, rolled it out to the launchpad, stood it up under the vast and empty Kazakh sky on its ungainly-looking skirt of engines . . . and then spent days, and more days, tinkering and adjusting and carefully pouring in its volatile fuel mix, and then carefully draining its volatile fuel mix, and staying up all night tinkering more, and chain-smoking awful Soviet cigarettes while Korolev shouted and blamed every glitch on Glushko, whom he suspected of designing flaws into his engines to sabotage the rocket and sink his career. Late at night some of the engineers would slink into

Voskresensky's cabin, he'd produce a bottle of cognac, which was strictly verboten yet utterly needed, and they'd commiserate.

When American rocketeers learned how big, heavy, and powerful the R-7 was, how many engines and combustion chambers and nozzles lifted it off the pad, they were bewildered. The Atlas, its closest American analogue, weighed only a third as much, and had only a third as much thrust at liftoff. How were the Soviets dwarfing American rockets? The truth was another close-held secret: the Soviets were years behind their competition in the development of strong, lightweight alloys, and in computers, and in the miniaturization of instruments. Everything was "heavy and bulky," Sergei Khrushchev admitted years later. Instruments were so unreliable that one or two backup systems were built in, adding to the weight. Also, to make up for their imprecise targeting, they packed as much explosive as possible into the warhead, adding yet more weight. Soviet rockets *had* to be so big and strong just to get themselves and their payloads off the pad. As an American scientist put it later, they "brute-forced everything."*

Multiple engines meant more possible glitches and problems. To lift off cleanly they all had to ignite at precisely the same time and put out the same amount of thrust. If one of them, for instance, failed to light and spilled liquid oxygen and kerosene, the flames roaring out of the other engines would ignite that fuel and cause a giant explosion on the pad. To light their engines, both Soviets and Americans used what were called pyrotechnic ignition

---

* Soviet aircraft were the same. At a time when the US was building lightweight, screamingly fast fighters out of titanium, the Soviets forged their MiG-25 out of nickel and steel. It was a flying anvil. To get in the air, it effectively used two cruise missiles for engines.

devices. The similarities stop there. The Americans, being Americans, spent wads of cash on high-tech electronic pyrotechnic devices. The Soviets, being Soviets, used what *Popular Mechanics* once described as "an oversized wooden match." When the R-7 was standing on the pad, techies scrambled under it and shoved broomstick-length rods of birch wood up into all of the combustion chambers. Small explosive charges on the ends of the sticks were all wired to a switch. When the switch was thrown it was hoped that all the charges would go off simultaneously and ignite all the engines simultaneously. It was classic Wrong Stuff: low-tech, but it worked often enough that it was still being used by the Russian space program decades after the Soviet Union disappeared.

A little before 9:00 p.m. on May 15 the techies shoved those sticks up the R-7's multiple orifices and ran for their lives, and the engines ignited and the rocket took off. Chertok was in a crowd of civilians and military men standing outside the bunker watching the rocket crawling into the night sky. Then, about a minute and a half into its ascent, its trajectory went, as he wrote, "lopsided." The rocket crashed into the desert 250 miles away.

In his biography of his father, Sergei Khrushchev tells a funny story he heard about this event. "Many generals had arrived from Moscow to watch the first launch of the R-7. It was impossible to fit them all in the command bunker." Korolev had a wooden reviewing stand built for them about a kilometer away. "A narrow dugout was located nearby—just in case. The generals watched as the rocket began to zigzag in the air. The four strap-on boosters separated from the main body like fireworks and flew in different directions." One of them "seemed to the observers to be coming directly at them, a fiery ball falling on them from the sky. This frightened even the battle-hardened generals. Shoving one another out of the way, they rushed toward the dugout. As they tried to squeeze through the narrow doors a brief scuffle broke out." The booster flew past and hit the ground with an explosion a safe distance away. The generals regained their composure. "During the days after the accident, soldiers picked up as souvenirs the shiny buttons with emblems torn off the generals' uniforms during the scuffle." (Apparently Sergei got it wrong and this happened at another failed R-7 launch, but it's still a good story.)

That first flight was a failure in another way: it alerted the Americans, who tracked it on radar, to the existence of a new launch site. A U-2 flyover soon brought back all the photographic evidence they needed. CIA analysts puzzled over what to call it. They had no assets on the ground in the Kyzylkum. Who would? Eventually one analyst found a tiny dot of a rail siding on a German Wehrmacht map from World War II. It said the place was called Tyuratam. The Cosmodrome had barely started operating and its cover was already blown.

The next R-7 tested exploded on ascent. The third, carrying a dummy warhead, seemed to be going fine when all guidance failed. It soared past its target site on the Kamchatka Peninsula and out over the Pacific. It was headed straight for the West Coast of the USA when it ran out of fuel and dropped into the ocean. Sergei says his father was relieved. He wanted the ICBM to help stave off World War III, not start it.

A test in August of that year went splendidly almost all the way. Ol' Number Seven delivered a dummy warhead four thousand miles from Tyuratam. Unfortunately, the nose cone that was supposed to protect the warhead melted away in the heat of the air friction, which would have been disastrous with a real warhead.

Still, it was the world's first ICBM. It just wasn't a very good one. And that was fine with Korolev. He was a space nerd, not an artillery officer. He was far more interested in reaching outer space than in pulverizing Manhattan. So, while some of his team worked on the nose cone problem, he rushed the rest to get a satellite up before the Americans did. His original idea for a satellite was much bigger and more complex than Sputnik. It would weigh more than three thousand pounds and carry many scientific instruments. But by the start of 1957 it was clear that the engines Glushko built for the R-7 couldn't yet throw that much weight into orbit, and some of the scientific equipment Korolev had hoped to stuff into the satellite wasn't ready anyway. His engineers told him that it could all be ready by 1958. Korolev didn't want to wait. He couldn't let the Americans beat him to it, after convincing Khrushchev he could get there first. Khrushchev had, of course, not the slightest idea or interest in what *kind* of satellite Korolev proposed to launch, as long as he launched it first. So Korolev drastically scaled the satellite down to what he called *Prosteishy Sputnik*—the Simplest Satellite. PS for short. Korolev's team nicknamed it SP for the boss, Sergei Pavlovich.

Chertok writes that he and other engineers at OKB-1 were "not at all excited" about working on Sputnik and merely "put up with Korolev's infatuation" with the thing. They thought solving the nose cone problem should be the team's sole focus. Korolev cleared some space at OKB-1 and put some of them to work on whipping out Sputnik as quickly as they could. With absolutely no plans, no technical drawings, and no time for either, "all the rules that had been in effect for the development of missile technology were abandoned," Chertok sniffed disapprovingly. "Almost all the parts were manufactured using sketches. Assembly wasn't conducted so much according to documents as according to the designers' instructions and on-the-spot fitting." In short, like so much of the Soviet space program, Sputnik was built quick, on the fly, by hand and ad hoc. No wonder it was so very simple, an aluminum sphere about the size of a beach ball with a shortwave transmitter and batteries inside, and four shortwave antennae outside. "Its internal electrical circuit was so basic that it would be a snap for any group of young hobby technicians to reproduce it." It only weighed about 180 pounds, most of which was the batteries. The Soviets were far behind the West and the Japanese in reducing the size and weight of batteries. Also, the interior of Sputnik was pressurized, because the Soviets had not yet figured out how to make electronics work in a vacuum.

At the start of October 1957, in the big assembly shed at the Cosmodrome, Sputnik was placed at the tip of a horizontal R-7, one of those nose cones was fitted over it, and the rocket was trundled by rail out to the launchpad, with Korolev and others in a solemn, worried procession beside it. Korolev had originally scheduled the launch for October 6, but he got spooked that the Americans were about to launch theirs and pushed his date up to October 4. His team scrambled. On the launch day they worked through glitch after glitch. Finally the R-7 blasted off near midnight. At 142 miles up, the second stage rocket engine cut off, the nose cone was jettisoned, and Sputnik floated free from the second stage of the rocket, which now trailed along behind it in orbit. Korolev agonized for the next ninety-six minutes until the satellite came back around the world to pass overhead. Chertok recalled "someone already quite hoarse was shouting, 'Everything is okay. It's beeping. The sphere is flying.'"

People around the world reported seeing Sputnik as a tiny, shiny dot streaking overhead in the morning or evening sky when the sun was low and lit it just right. They did not. Sputnik was much too small to be seen going by. But some did see *something*—the R-7's second stage, following along behind Sputnik. Displaying his innate showmanship again, Korolev had cleverly covered it in a shiny surface and buffed it to a mirror finish so that it would gleam in the sun. It could be seen through binoculars, or without them if your eye was very sharp.

Anyone with a shortwave radio, which were very popular, could hear Sputnik go by. *Radio*, a Soviet magazine, helpfully published the frequency, 20.005 MHz, along with what would be the satellite's orbital characteristics. People around the world tuned into Sputnik's quick, chirpy *beep beep beep* as it passed over their heads. *Life* described it as an "eerie, intermittent croak . . . like a cricket with a cold," which may say more about *Life* than about Sputnik. You can hear it online.

American rocketeers were chagrined. They had been racing the Russians to put the first satellite up. Their scrappy competitors had beaten them to it. And, ironically, the Americans had only themselves to blame for goading the Soviets into it. Flying U-2 spy planes over the Soviets' heads was risky business, a clear and provocative violation of international law regarding sovereign air space. But could the same be said of orbiting satellites? Where did air space end and outer space begin? In 1955 the Eisenhower administration announced that in honor of the International Geophysical Year (IGY)—actually July 1957 through December 1958, planned by scientists around the world as a period of open cooperation on various earth sciences projects—the US would try to launch a civilian, scientific satellite. Four days later, the Soviets announced they would, too. The race was on. Behind it was the presumption on both sides that once a civilian satellite had established the precedent of "freedom of space," they could send up all the military and spy satellites they liked. It was Cold War gamesmanship par excellence.

Now the army, navy, and air force vied for the honor of putting the first satellite up. For "honor," read "funding." Coming out of World War II, when the money spigot had been turned on full, the armed services had found themselves in the postwar years competing for reduced support. In rocketry, the US Army had von Braun and his V-2s, while the navy had their own ideas about how to develop ICBMs. When the air force was split off from the army in 1947, they had little interest in missiles at first. Their top brass argued that winged aircraft flown by human pilots were superior—i.e., Curtis LeMay's Strategic Air Command bombers. The Soviets' successful testing of their own hydrogen bomb in 1953 was a wake-up call. The air force began to take ICBMs more seriously, though they continued to refer to missiles as "unmanned aircraft" for a while. The Eisenhower administration that came into power that year calculated that a nuclear war would be an air war fought in minutes with ballistic weapons. The race to develop ICBMs was on—a race not just against the Soviets, but among the three competing branches of the US military, all with their own missile programs, all jostling for support in Washington. Von Braun's army team worked on the Redstone, the air force had the Minuteman and Atlas, and the navy had the Vanguard and the submarine-launched Polaris. As they lobbied and elbowed one another, Eisenhower chastised them more than once for bickering like squabbling siblings. But it seemed they couldn't help themselves. Even as the chair of the Joint Chiefs of Staff, Air Force General Nathan Twining argued that the army's missiles should be restricted to a two-hundred-mile maximum range, making them in effect a new sort of battlefield artillery, while leaving the longer-range stuff to his air force, better equipped to handle things that fly.

Von Braun and his army team begged to differ. They announced in 1956 that they were ready to launch a Redstone rocket with a small satellite they called Explorer I on top. But in the end they, and the air force, were out-lobbied by the navy, who won Eisenhower's approval for the honor of sending up the first US satellite on its Vanguard rocket. The Vanguard was a complicated machine that was going to take far longer to develop than von Braun's, but the betting in Washington was that it would still beat the Soviets. In keeping with Eisenhower's IGY pronouncement, Vanguard was reorganized as an ostensibly civilian project, but the US Navy uniforms were clearly visible behind that fig leaf. The Vanguard was still inching painstakingly toward readiness when Sputnik began broadcasting its *beep beep beep*.

★ ★ ★

The average American wasn't particularly impressed or vexed by Sputnik. For Americans in 1957, threats from outer space meant flying saucers and aliens with ray guns, not a circling beachball that couldn't do anything but go *beep*. Game three of the World Series (Braves v. Yankees) the next day caught more attention—it was the first World Series game ever played in Milwaukee. Sputnik was little more than fodder for pop culture references, like the cutesy love song "Satellite" by Teresa ("I Saw Mommy Kissing Santa Claus") Brewer, torturing metaphors about how her love was like a satellite "in an orbit around your heart."

Roosevelt Sykes's "Sputnik Baby" and Skip Stanley's "Satellite Baby" offered more rock and roll versions of the same sentiment. There were joke songs like Buchanan and Goodman's "Santa and the Satellite." Jerry Engler's "Sputnik (Satellite Girl)" fell somewhere between rockin' and jokin' with its goofy-catchy chorus about a "spoo-spoo-spoot-a-nick-a-chick." (The band that outdid them all, however, was Swedish. A Ventures-style guitar group calling themselves the Spotnicks appeared in Gothenburg almost as soon as Sputnik appeared in the sky. They wore spaceman outfits and played catchy tunes with "out-a space" or Russian themes. They were big in Europe and England for a few years but, tellingly, unknown in the US.)

Meanwhile, the Starlite Hotel near Cape Canaveral got new space-themed murals and advertised itself as "Sputnikked up." Bartenders around the country started serving the Sputnik cocktail, made of course with vodka. By 1958, the *-nik* from Sputnik had been borrowed for the diminutive term for certain American hipsters—beatnik.

No, Americans weren't much concerned about Sputnik—until politicians and the press told them they should be. Democrats seized on Sputnik as a stick to beat Eisenhower and the Republicans with. The venerable Speaker of the House Sam Rayburn declared that "the people of the United States have been humiliated. They're disturbed, and they're unhappy." Minority whip and future presidential candidate Hubert Humphrey opined that "we certainly don't want to become hysterical," but "let's face the fact that we've taken a licking . . . and it has embarrassed us throughout the world." The news media backed them up, identifying it as a "space race" for the first time—and already declaring the Soviets the winners. As early as November *The Economist*, not known for going out on any limbs, predicted that the first human to set foot on the Moon "is almost certain to be a Russian." *Life* declared that without a major course correction "it is reasonable to expect that by no later than 1975 the United States will be a member of the Union of Soviet Socialist Republics." The military and defense contractors smelled funding increases and chimed in.

Two Democrats rode Khrushchev's rockets all the way to the White House. In 1958, gearing up for a 1960 presidential run, Massachusetts senator John F. Kennedy unleashed the term "missile gap," claiming that Eisenhower had allowed the Soviets to build up a superior force of ICBMs that could annihilate "85 percent of our industry, 43 of our 50 largest cities, and most of the nation's population." He repeated those figures over the next two years, clearly unperturbed that there wasn't a shred of evidence to back them up. Politics is politics. His running mate Lyndon Johnson went full-on Flash Gordon: "Control of space means control of the world. From space the masters of infinity would have the power to control the earth's weather, to cause drought and flood, to change the tides and raise the levels of the seas, to divert the gulf stream and change temperate climates to frigid." Picture each of those sentences ending with an exclamation point or two and they could be spoken by Ming the Merciless.

As Eisenhower's vice president, Kennedy's opponent Richard Nixon was privy to classified intelligence proving that the missile gap was a lie. But he couldn't go public with that top secret information, "and Kennedy, the bastard, knows I can't," he once growled.

Khrushchev helpfully chimed in, bluffing and exaggerating. It has been called "The Bluff of the Century." Already on October 10 he told the *New York Times* that "we have all the rockets we need: long-range rockets, intermediate-range rockets and close-range rockets." In November he claimed, "We have developed an intercontinental ballistic missile with a hydrogen warhead." It was almost entirely hooey, but who was going to stop him from spouting it? Certainly not the ambitious Kennedy. In effect, Kennedy was partnering with Khrushchev in propping up this front of a missile-mighty Soviet Union. Call it a duet of doom. Both of them benefited politically from the lie: Kennedy rode it to the White House, and Khrushchev used it to keep his enemies, domestic and foreign, at bay.

In truth, Khrushchev needed the triumph of Sputnik more than he could ever have guessed back in 1956 when he approved it with a shrug. Up until Sputnik, 1957 had been a rough year for him. In June, a group of twelve old-line Stalinists opposed to some of Khrushchev's attempted reforms, and his closed-doors denunciation of Stalinism in 1956, tried to depose him. They included his former partners Malenkov and Molotov, for whom the fiery "cocktail" was named.* During a long shouting match behind closed doors in the Kremlin, with the conspirators accusing Khrushchev of everything from betraying Marxism to minor personality flaws, his lackey Leonid Brezhnev stood up to defend him, but was so nervous he fainted. Khrushchev's real defender was Marshal Georgy Zhukov, the most popular hero in the USSR, the general with the pounds of medals slathered across his chest who had driven the Germans out of Russia and all the way back to Berlin during the Great Patriotic War. He was adored by the Soviet people and, more to the point, by the entire Red Army. That had made him the one man in all of the USSR whom Stalin could not make a move against, although he did bundle him off to the Ukrainian port of Odessa to get him out of the way. Bringing him back to Moscow had been one of Khrushchev's first moves when Stalin died. In return, Zhukov backed Khrushchev now. Khrushchev survived the coup. No one doubted what Stalin would have done to the conspirators, but Khrushchev showed he was not his predecessor: he merely had them stripped of their rank and privileges and posted to the boonies. His punishment for Molotov, the most internationally sophisticated of the bunch, was to make him ambassador to Outer Mongolia. There were grins and snickers all over the Kremlin.

But then Khrushchev showed he *was* like Stalin in one respect: paranoia. In the months following the failed coup, Khrushchev

---

* Finnish freedom fighters desperately defending their country against Soviet invasion forces in the Winter War of 1941 threw homemade petrol bombs at Soviet tanks. Molotov, the Soviet foreign minister, outrageously claimed that Soviet bombers were not dropping ordnance on the Finns, but food parcels. The Finns sarcastically nicknamed their petrol bombs Molotov cocktails, a drink to wash down the fictious food.

became convinced that Zhukov, drunk with success, was plotting his own move against him. Khrushchev's fears were not entirely unfounded. His military leaders were unhappy—and the missile program was at the heart of their discontent. "Since starting work on the ICBM, Khrushchev had unilaterally slashed troop forces by a staggering 2 million men," Brzezinski writes. He had reduced conventional artillery and armored divisions, closed bomber factories, and canceled the building of naval ships. He was *restructuring the entire Soviet military* around Korolev's promise of ICBMs. Understandably, the officer corps was grumbling. Khrushchev began to fear that Zhukov wanted to be another Eisenhower, an old warhorse turned political leader. He had to be removed from power—but, because he was the greatest hero in the land, it had to be done quietly and gently. Early in October 1957, Khrushchev arranged for Zhukov to be on a navy ship in the Adriatic, heading to a goodwill tour of Yugoslavia and Albania. Meanwhile Khrushchev slipped off to Kyiv. While they were both far from Moscow, a tiny item appeared on the back page of *Pravda*, announcing that Zhukov had been relieved of his duties. Khrushchev was in Kyiv so as not to appear to have anything to do with it. He was delighted over the next few days that Sputnik took over the front pages of newspapers worldwide. (Zhukov went on to enjoy a quiet retirement, did a lot of fishing with tackle sent to him by his fellow old warhorse Dwight Eisenhower, and was given a hero's funeral when he died in 1974.)

Khrushchev was so pleased with Sputnik that he asked Korolev to launch something even bigger and better into orbit for the November

7 anniversary of the Revolution—*three weeks away.* Thinking on his feet, as dealing with Khrushchev often required, Korolev said he could put a dog into orbit next. Khrushchev loved it.

Now all Korolev had to do was make it happen. In three weeks. His team had only a few sketches for a follow-up to Sputnik. Not for the last time, they scrambled and improvised madly. They had another Sputnik they'd built for ground tests. They retrofitted that with a harness for a small dog. To make the thing bigger than Sputnik, they'd simply leave the last stage of the launch rocket attached.

Putting dogs on top of rockets was nothing new. Since so little was known about the effects that blasting off in a rocket might have on the human body and brain—the g-force of acceleration, the disorientation of weightlessness, the impact of radiation, the g-force of deceleration—the Soviets and the Americans both had been using various species of animals to test conditions since the 1940s. The Americans started sending up fruit flies aboard their White Sands V-2s in 1947. An anesthetized rhesus monkey they named Albert II (not to be confused with Albert II, Prince of Monaco, who resembled a monkey only as much as we all do) went up eighty-three miles in a V-2 in 1949. Unfortunately, his parachute failed to open on reentry and he was smashed to death on impact with the ground. The Americans continued to send up primates in the 1940s and 1950s. Something like two-thirds of them died. They used many other species as well, maybe the oddest of which was black bears, who were strapped into a rocket-powered sled at a facility with the deceptively sweet name the Daisy Track to test the physical effects of ultra-rapid acceleration and deceleration.

The French space program, such as it was, preferred cats. They yanked fourteen strays off the streets of Paris and trained them.

They chose a female they designated C 341, whom French journal-ists called Félicette, feminine diminutive of Felix. They put her in a space harness, attached electrodes to her brain, and launched her on a V-2–based rocket from their facility in Algeria. After reaching 157 kilometers and experiencing a few moments of weightlessness, Félicette parachuted to a soft landing in the Sahara, unharmed. Two months later scientists euthanized her so they could perform an autopsy. After decades of obscurity, Félicette was honored in 2020 with a bronze statue funded by a Kickstarter campaign. The sculptor depicted her sitting on top of the world, a manifest exercise of artis-tic license.

The Soviet program followed the advice of the famous Pav-lov Institute of Physiology: "Dogs are highly organized, steady and easily trained animals. Monkeys on the other hand are capricious, highly strung and undisciplined." Zoologists captured strays on the streets of Moscow. They preferred mutts because they believed them more emotionally stable than purebreds. At the Cosmic Physiologi-cal Research Station the mutts spent two months experiencing con-ditions like they would in flight, from loud noises and high heat to long periods of tight confinement. The first dogsmonaut pair went up in 1951; they survived. By 1957 the Soviets had sent dozens up on suborbital flights, some dogs more than once, with a much higher survival rate than the Americans' poor primates.

Sputnik 2, however, was to be the first time an animal went into orbit. Korolev's team selected two dogs from their stock of test canines. One was a small, docile spitz-husky mongrel they officially named Kudryavka (Little Curly). Because she was noisy the team nicknamed her Laika, Barky, which is how she came to be known to the world. The American press would call her Muttnik or Mutnik.

Laika's backup crew was Albina, Whitey. Albina was a veteran dogs-monaut with two suborbital flights behind her. But because she had recently given birth to puppies, the team didn't have the heart to send her up. They knew this was a one-way trip. Korolev and his team could put an object into orbit, but they had no idea yet *how to bring it back*. Laika was going to be a martyr to Khrushchev's compulsion to embarrass the Americans.

There was enough oxygen on board for a week. Rather than have Laika die horribly of asphyxia, they rigged a lethal injection system they could trigger from the ground when the time came. On November 3 they strapped her into her harness and hooked her up to electrodes to monitor her heart rate and breathing. The R-7 carrying Sputnik 2 successfully blasted off. The crushing pressure and deafening noise of liftoff clearly terrified Laika; the monitors showed that her heart rate shot up three times and her breathing was twice as fast as normal. Worse, the heat shielding in her tiny space failed. She was stressed, overheated, and experiencing the disorientation of weightlessness for the first time. Nine hours into her flight her monitors flatlined.

At first, characteristically, the Soviets issued false reports that she was doing fine. Nine days later, they reported that she had died quietly from a lack of oxygen after seven days in flight. That was not only false, it was not much comfort to dog lovers around the world, who pictured Sputnik 2 passing over their heads carrying Laika's corpse every 104 minutes for the next five months. *Time* called it "a half-ton tomb for a dead dog." The magazine did not mention that on the same day Laika was rocketed into space, a black bear named Oscar was rocketed horizontally at the Daisy Track. He survived, but then, like Félicette, he was murdered so he could be autopsied.

On April 14, 1958, Laika was finally cremated as Sputnik 2 burned up reentering the atmosphere. It wasn't until 2002 that the true conditions of her death finally came out.

Laika may have died pitiably, but in Soviet propaganda, and in world culture, she lived on as an icon, the very embodiment of man's best friend, a canine hero of the people who sacrificed her life for the good of all, whether she knew that or not. There were Laika brand cigarettes; Laika postage stamps; songs about her; a graphic novel about her; children's books about her; Mighty Sparrow's calypso ode to her, "Russian Satellite"; the movie *My Life as a Dog*; a stop-motion animation studio named for her; the Laika-inspired Soviet dogs-monaut in *Guardians of the Galaxy 3*; and a crater on Mars named for her.

The Americans scrambled. On December 6, on a live network broadcast from Cape Canaveral, the navy and its civilian part-ners tried to launch their Vanguard rocket with a grapefruit-size, three-pound satellite on top. The rocket's engines roared, it rose three feet off the pad, then millions of Americans gasped as it sagged back down and blew up in a huge fireball. When this hap-pened in the Soviet program, they would joke that the missile had "returned to base." (But they would *never* televise it.) The rising fireball glowed a Halloweenish orange, but Americans didn't see that on their black-and-white TVs. Vanguard used a liquid oxygen and kerosene fuel, which if nothing else blew up quite spectacu-larly. The pitiful little satellite popped off the top like a cork and arced seventy-five feet to drop onto the beach nearby. One member of the press said its audible beeping was a pitiful sound, as though it was a wounded animal that someone should put out of its misery. A technician did.

"OH! WHAT A FLOPNIK!" the *Herald* of London cried. The *Daily Express* went with "Kaputnik."

In January von Braun's team got their Explorer I into orbit. At only thirty pounds and six inches in diameter it was puny compared to the Sputniks, and Khrushchev had a field day mocking it. *Life* quoted an unnamed "Muscovite" who scoffed that "Americans design better automobile tailfins but we design the best intercontinental ballistic missiles and earth satellites." But that was a rare note of skepticism in the American press. Korolev, obsessed with his rival, watched with interest as von Braun was hailed as a great American hero. He was on the cover of *Time* with a giant rocket lifting off in flaming glory behind him. The *New York Times* and the rest of the media went on and on about his genius. The media acted as if the Sputniks had never happened and von Braun's bauble was the most remarkable achievement of humankind in all history. They acted as though von Braun's Nazi and SS past and use of slave labor had never happened either, leaving it for the musical satirist Tom Lehrer to mention the obvious in song:

> *A man whose allegiance*
> *Is ruled by expedience*
> *Call him a Nazi, he won't even frown*
> *"Nazi, Schmazi!" says Wernher von Braun.*

According to those who were around him, the anonymous Chief Designer didn't begrudge von Braun his fame. He just envied his budget.

On November 4, the evening after Sputnik 2 was launched, Eisenhower had the top brass from all the armed services to dinner at the White House. He accused them, accurately, of being "more interested in the struggle with each other than against an outside foe," retarding American missile development while the Soviets leapt ahead twice in less than a month. It had to end. The top brass eyed their rivals around the table, muttering and grumbling. Eisenhower soon signed a bill to create a central, ostensibly civilian, space agency, NASA. At the same time, he authorized the start of a separate agency, ARPA, the Advanced Research Projects Agency (later DARPA, with Defense added in front), to oversee coordinated military space activities, usually top secret.

The cover of the January 6, 1958, issue of *Time* revealed their Man of the Year for 1957: Nikita Khrushchev. He was depicted on the cover grinning like a goblin, wearing a Kremlin-shaped crown and bobbing Sputnik I in his pudgy hands. He was described inside as "stubby and bald, garrulous and brilliant." *Time*'s editors opined: "With the Sputniks, Russia took man to a new era of space, and with its advances in the art of missilery, posed the U.S. with the most dramatic military threat it had ever faced. And with the Vanguard's witlessly ballyhooed crash at Cape Canaveral went the U.S.'s long-held tenet that anything Communism's driven men could do, free men could do better. Whatever the future might bring, in 1957 the U.S. had been challenged and bested in the very area of technological achievement that had made it the world's greatest power."

It was one of the proudest moments of Khrushchev's life.

5

KHRUSHCHEV'S CHARM OFFENSIVE

We were hiding only one thing—that there was nothing
to hide.

—SERGEI KHRUSHCHEV

Long before President Kennedy said anything about it, reaching
the Moon (and then other planets) was a goal cherished both by
Sergei Korolev and Wernher von Braun. Both sides started launching lunar probes in 1958. The Americans went from one dismal
failure to another in 1958 and 1959. One of them, launched in 1959,

managed to get within 37,000 miles of the Moon, but most of the others never even got out of Earth's atmosphere.

Korolev didn't fare much better at the start. He fondly nick-named his small Luna probe Mechta, Dream. It was a dream deferred at first. His initial two attempts in 1958 blew up shortly after takeoff. In typical Soviet fashion, these failures were not made public. When the third attempt, in January 1959, appeared to be on its way to the Moon, the Soviets announced the successful launch of "Luna 1." It was a failure in its own way—it missed the Moon by 6,000 kilome-ters. The Soviets, compounding their subterfuge, claimed that it was *planned* as a flyby. Khrushchev jawboned that it was "a new exploit of worldwide importance, having successfully launched a multi-stage rocket *in the direction of* the Moon." [Emphasis added.] "Even the enemies of socialism have been forced . . . to admit that this is one of the greatest achievements of the cosmic era." Another Moon shot that June failed, and was covered up. After a few months of tinker-ing, Korolev launched "Luna 2" on September 12. His *mechta* was finally fulfilled. It landed on the Moon on September 14, 1959, the first man-made object to do so.

Nikita Khrushchev landed in America the next day, the first Soviet leader to do so. He stepped down out of a giant Aeroflot Tu-114 airliner and marched across the tarmac at Andrews Air Force Base trailing his sturdy common-law wife Nina and other stout females, grinning like a happy, prospering butcher taking the fam out to dinner.

Arriving in the Tu-114 was an opening bluff. Attending a con-ference in Geneva in 1955, Khrushchev had been mortified that his dinky two-engine Ilyushin Il-14, modeled on the American DC-3, was half the size of the sleek, elongated Lockheed Super

Constellation that brought Eisenhower. His son Sergei remembered that the Ilyushin "looked like an insect" next to Eisenhower's stretch limo of the skies. Back in Moscow, Khrushchev had ordered up the biggest airliner Soviet designers could produce. Rather than build a new one from scratch, they simply converted a Tupolev Tu-95 turboprop long-range bomber and called it the Tu-114. Tupolev had already done this once before, converting a bomber into the Tu-104, the world's first passenger jet. It was infamously unsafe. A Soviet joke went: "Tu-104 is the fastest plane in the world. In just five minutes it will carry you to the grave." His supersonic Tu-22 bomber was even scarier. Rushed into production with a typical Soviet attitude of fly-it-now-fix-it-later, it killed so many crews that they called it the Maneater and the Errorplane. The only thing pilots liked about it was that it used an alcohol coolant, which they drained off and drank to toast their luck if they survived a flight.

Tupolev debuted the Tu-114 on November 3, 1957, the very day Laika was launched into the history books. The largest, fastest, longest-range airliner in the world, the Tu-114 suggested that the Soviet Union was well ahead of the West in air travel as well as space travel. What was carefully hidden from the Westerners was the fact that this one was the only Tu-114 in operation at the time. And it should not have been. Before it took off in Moscow, technicians detected serious problems with its turboprop engines. Advisors begged Khrushchev to take another plane, but he decided it was worth the risk to himself, and everyone else in the plane, to wow the West. He later confessed in his memoirs that he was terrified all the way across the Atlantic. Then again, flying in *any* Aeroflot aircraft in those days was a crapshoot. Westerners who had cause to do so were appalled at how rickety the aircraft were and how devil-may-care the crews seemed about basic safety procedures like not taking off unless all the engines were running and not using hand-drawn maps to find their destinations.

Khrushchev had eagerly accepted President Eisenhower's invitation to come visit America. Eisenhower hoped that face-to-face he might learn more about his mysterious nuclear counterpart before the two of them accidentally blew up the world. He knew so little about Khrushchev that he sometimes mispronounced Nikita as "Nikito." Khrushchev was just as ignorant about the US. When told that Eisenhower wanted them to spend a few days at Camp David, he replied suspiciously, *"Kemp David?* What sort of camp is it?" He hadn't evaded getting put in any of Stalin and Beria's camps to fly halfway across the world and get put in another. He also wanted to ease tensions—and he had more reason, given how much his claims of Soviet nuclear might were fiction and bluff. His greatest worry was how vulnerable the USSR was to an American preemptive strike. His only resort was to keep the Americans guessing with a balancing act of belligerent bluster and disarming charm.

Crowds lined the route from the air base to the White House. They were strange crowds—not welcoming, rather sullen, really. They didn't wave, didn't shout; they just stared. They were as mystified by the Soviet leader as their president was—but manifestly less trusting. Hiding behind what Churchill had called its Iron Curtain, the Soviet Union in 1959 was still struggling to overcome the ravages of the war and Stalin's madness. The America that Khrushchev was visiting, on the other hand, was as close to a capitalist paradise as the world had ever known. Despite lingering groups stuck in poverty and disenfranchisement, the majority of Americans were prosperous, comfortable, and awash with consumer goods—new homes, giant cars, washer-dryers, poodle skirts, bullet bras, bicycles, console TVs (mostly still black-and-white). Popular music was dominated by the soft, soothing fluff of Fabian, Connie Stevens, and Frankie Avalon. Elvis and his wiggling hips had been shoved out of sight in the army; the sexy menace of rock and roll and "race music" had been tamed. The top TV shows were Westerns—*Gunsmoke*, *Wagon Train*, *Have Gun—Will Travel*—with choreographed shootouts but never a drop of blood.

What ruffled Americans' complacency in 1959 was this man with his threats of nuclear annihilation. Some Americans dug useless bomb shelters in their backyards. Their kids learned the equally useless "duck and cover" maneuver in school. You couldn't go far in any city without seeing a black-and-yellow fallout shelter sign.

America in 1959 was also very much One Nation Under God—the phrase had just been added to the Pledge of Allegiance in 1954. Many believed that Soviet atheism was as threatening as Soviet nukes. On the eve of Khrushchev's visit, New York's deeply conservative Cardinal Francis Spellman explicitly warned that the Soviet leader would bring "propaganda more lethal than explosives." The bishop of Paterson, New Jersey, arranged for Masses offered for the victims of "atheistic communism."

And now here was this leader of the godless Communist world, the embodiment of the Red Menace . . . a short, round, bald Antichrist rolling past them in a limo, grinning, wagging stubby fingers at them. No wonder most in the crowd had no idea how to react. (A few did. A refugee from Soviet-dominated Poland stood outside the White House on a hunger strike to protest the visit. And a small group who called themselves the Committee for National

Mourning waved skull-and-crossbones pennants at the passing motorcade.)

Khrushchev's mischievous impulse to tweak the capitalists was on full view right away in the Oval Office, where he beamed and presented President Eisenhower with a gift. Luna 2 hadn't landed on the Moon so much as crashed into it. A soft lunar landing was still years off. When Luna 2 hit the surface, it smashed open and two palm-size steel balls, one three inches in diameter and the other five, spilled out. The surface of each was made up of seventy-two pentagonal segments fitted together, some with the hammer and sickle embossed on them, others with CCCP, Cyrillic for USSR. These balls burst on impact, scattering the 144 little souvenirs, or "pennants," as the Soviets said, all around on the dusty surface. In this sense, the Soviets "planted their flag" on the Moon a full decade before American astronauts did theirs. The site of the first man-made landing on the Moon was also the site of the first trash dump on the Moon.

In the Oval Office Khrushchev grinned like a Slavic Cheshire cat as he handed Eisenhower a replica of one of these balls. Then, through his interpreter, he outright trolled the president about it: "We have no doubt that the excellent engineers and workers of the United States of America who are engaged in the field of conquering the cosmos will also carry their pennant over to the Moon. The Soviet pennant, as an old resident, will then welcome your pennant and they will live there together in peace and friendship."

At the start of a "goodwill" visit it was an act of appalling cheek. The Soviet leader couldn't have looked more pleased with himself if he'd handed the president an actual slice of humble pie. While Eisenhower managed a thin, pained smile, Vice President Richard

Nixon looked ready to throttle the gloating Commie boor. It was probably best for all concerned that Korolev didn't get "Luna 3" off the ground until October 4, when Khrushchev was back in Moscow. It swung around the Moon and took photos of the far side, the first time human eyes had ever seen it. The Americans would keep failing to hit the Moon until 1964.

Over the next two weeks, Khrushchev, his family, and attendants crossed the continent, hitting both coasts—Washington and New York City in the East, Los Angeles and San Francisco in the West, with a Grain Belt stop in Iowa and a Steel Belt visit to Pittsburgh. Newspapers and television covered every minute of it; a vast cloud of more than three hundred reporters traveled with him. Khrushchev understood the power of the media and worked ceaselessly to win over the Americans. For a few days he was on Americans' black-and-white TV screens more than Lassie and Matt Dillon combined, with running commentary by national news figures like the sepulchral David Brinkley and the bow-tied, impossibly soft-spoken Dave Garroway. One newsman with delusions of Twain waxed sarcastically elegiac about Khrushchev's "spraying the air with the perfume from his atomizer of charm." The whole affair has been summed up as the Cold War Roadshow.

On a train from Washington to New York, Khrushchev engaged Henry Cabot Lodge Jr., the tough-talking US ambassador to the UN, in a friendly debate about which side would win a nuclear war. It was a favorite topic of Khrushchev's, reinforcing the image of the US and USSR as coequal nuclear superpowers, his grandest and most persistent bluff. New Yorkers also lined the streets as his motorcade rushed by, many of them holding up handmade signs; one read YOU

CAN REACH FOR THE MOON BUT YOU CAN'T GET AMER-
ICA. He visited the Empire State Building and shrugged. "If you've
seen one skyscraper you've seen them all."

When Khrushchev flew to Los Angeles he went straight to the
Twentieth Century Fox studios, where he watched them shoot a
dance scene for *Can-Can* that he later called exploitative and por-
nographic, especially the part where a male dancer slid under Shirley
MacLaine's flouncy skirts and appeared to emerge with her undies
in his hands. The Khrushchevs had a big luncheon with Hollywood
royalty, carried live on television. It was widely noted that Marilyn
Monroe, notoriously late for everything, arrived early. The famously
anti-Communist actor Ronald Reagan refused to attend.

During the lengthy affair, the Los Angeles police chief told
Lodge that he could not guarantee Khrushchev's safety during a
planned trip to Disneyland, which Nina Khrushchev was especially
looking forward to. Peter Carlson, who wrote an engaging book
about Khrushchev's visit, *K Blows Top*, recounted how Khrushchev
pounced:

> "Just now, I was told that I could not go to Disneyland," he
> announced. "I asked, 'Why not? What is it? Do you have
> rocket-launching pads there?'" The audience laughed.
>
> "Just listen," he said. "Just listen to what I was told:
> 'We—which means the American authorities—cannot guar-
> antee your security there.'"
>
> He raised his hands in a vaudevillian shrug. That got
> another laugh.
>
> "What is it? Is there an epidemic of cholera there? Have
> gangsters taken hold of the place? Your policemen are so

tough they can lift a bull by the horns. Surely they can restore order if there are any gangsters around. I say, 'I would very much like to see Disneyland.' They say, 'We cannot guarantee your security.' Then what must I do, commit suicide?"

Frank Sinatra, who was sitting at the table with Nina, said he'd squire the Khrushchevs there himself and personally guaranteed no one would mess with them. They did not take him up on the offer.

From LA the Khrushchevs took a train up the coast. In San Jose he visited an IBM plant making computers and shrugged again, declaring that Soviet computers were superior. (They were not. The Soviets were in fact hopelessly behind the Americans.) But he was genuinely impressed with the plenty and variety of foods in the plant's cafeteria. The process of sliding his tray along the line and picking any foods he liked was something he confessed one did not experience in his homeland. He was also struck with supermarket envy during the trip. As he wryly put it, the Soviet people were better at producing moon rockets, but the Americans were maybe better at making sausages. After he returned to Moscow, Soviet-style cafeterias bloomed in workplaces, and a few supermarkets appeared in Leningrad and Moscow, though both tended to be poorly stocked, gloomy imitations of what he saw in America.

During this second half of his trip the former farm boy and factory worker insisted on getting out from behind the heavy security detail that surrounded him. FBI director J. Edgar Hoover had estimated that as many as twenty-five thousand Americans might try to assassinate him, a figure he apparently pulled out of the air. Khrushchev didn't care. He wanted to meet real American workers. His politician's instinct proved accurate. The more American hands he shook, the more babies he hugged, the more hot dogs he ate, the more English he tried to speak ("very okay" became a favorite sort-of-American phrase, delivered with a pudgy thumbs-up and "very" pronounced "wery"), the more Americans warmed to him. Gradually, when he grinned and waved at them, they grinned and waved back. They laughed when he patted a man's beer belly and declared him "a real American." In a Pittsburgh factory a worker handed him a cigar. Khrushchev took off his wristwatch and gave it to the man. A few days later the man took it to a watch shop and asked its value. He was crestfallen to hear it was only worth twenty dollars.

Khrushchev seemed happiest of all in a cornfield near Coon Rapids, Iowa, on the farm of the one man in America he could sort of call a friend, Roswell Garst. Since the 1930s, farmer Garst had been evangelizing a hungry world about the benefits of hybrid seed corn. Khrushchev, with millions of hungry mouths to feed, became a true believer in 1955—people joked that he went *kukuruznik*, more or less "cuckoo for corn"—when he brought Garst over for a visit and an unlikely bromance ensued. Soon many millions of acres of Soviet "virgin lands" would be planted with Garst's corn. The very disappointing yields—they were virgin lands for a reason—would contribute to Khrushchev's eventual downfall; his successor Leonid Brezhnev would have to stoop to the humiliation of buying immense stores of wheat and, yes, corn from the Americans. But now Khrushchev was delighted to meet up with Garst again. They strolled the cornfield, happily chattering about grain through an interpreter, pausing once to yell at the huge crowd of newsmen, in English and Russian, for trampling some of Garst's plants.

Khrushchev could still turn petulant when having to duck pointed questions from the press such as, for example, his plans for Berlin, divided into Soviet-controlled and Allied sectors. (He loved baiting the West about Berlin. He told Sergei that Berlin was the West's balls, and he liked squeezing them. Two years later he would get more serious and build the Berlin Wall.) But the tantrums came less often in the last few days of the trip. The media, like the public, had been won over by the jolly, roly-poly Russki, very different from the baby-eating monster they'd expected. From his Tu-114 heading back to Moscow on September 27, he radioed his thanks to the president and the American people, and his conviction that the trip "will definitely help to ease international tension, to strengthen the cause of universal peace."

That very October, the prestigious journal *Foreign Affairs* published a long essay, "On Peaceful Coexistence," by Nikita S. Khrushchev, written well before his trip. "We say to the leaders of the capitalist states: Let us try out in practice whose system is better, let us compete without war," he proposed. "This is much better than competing in who will produce more arms and who will smash whom."

So there was a whiff of hope that the Cold War might indeed be thawing. It lasted seven months, until May 1, 1960, when the Soviets shot down Francis Gary Powers's U-2 spy plane.

The CIA's high-altitude reconnaissance planes had been flying out of air bases in Europe and the Middle East, violating Soviet air space with impunity, since 1956. This mortified and aggravated Khrushchev no end. But like some other things that vexed him, it was at least partly his own fault. At that Geneva conference back in

1955, President Eisenhower, citing the threat a nuclear arms race posed to both sides, had proposed a program called Open Skies, by which the US and USSR would allow each other to fly freely over their territories and photograph any suspected nuclear facilities or missile launch sites they spotted. Khrushchev rejected the offer for the disingenuous, one-sided scam it was: while US planes could easily fly over Soviet territory from those air bases that ringed the Iron Curtain, the Soviets had no comparable bases near the US, and no plane had the range to fly from Soviet territory to the US and back again.

A year later, in a grand show of international friendship and openness, Khrushchev invited representatives of foreign nations, including the US, to attend a big air show at Moscow's Tushino airfield. Nine US Air Force generals gladly took him up on the offer. One of them was General Nathan Twining, chair of the Joint Chiefs of Staff. Another was Curtis LeMay. Khrushchev wanted to impress on these men that he had a mighty heavy bomber force too. It was one of his most desperate bluffs yet. As his son Sergei relates, "Soviet heavy bombers could be counted on your fingers: about ten 3Ms [a jet-powered strategic bomber known as the Bison] and three or four TU-95s [giant but propeller-driven]." So the Soviet pilots used an ancient trick: The bombers "that flew sedately over Tushino in formation of threes had hardly disappeared from view when, after gunning their motors and circling rapidly over Moscow, they reappeared over the grandstands. Over and over again."

LeMay was convinced that day that there was a "bomber gap" between the military technologies of the Americans and the Soviets. There was, but it was overwhelmingly in his favor. Still, when he returned to Washington he would convince Eisenhower that more B-52s were needed right away to match the Russki squadrons he thought he saw rumble overhead that day. At a reception afterward, Khrushchev, red-faced and tipsy, leaned over a banquet table and said to General Twining, "We have very good missiles. We'll show them to you if you show us yours." The dumbfounded Twining didn't know how to respond. Khrushchev mistook Twining's surprised silence for arrogance. So a few days later he upped the ante and announced that the USSR had ICBMs that could travel eight thousand miles.

This too was a lie, and like some of his others, it backfired on him. The Americans now decided they needed to see for themselves—to peek through the keyhole into their neighbors' bedroom, as Khrushchev grumbled. (With an appropriate sense of irony, the CIA would give a satellite program for spying on Soviet territory the code name KEYHOLE.) Eisenhower had been reluctant until then to authorize flights of the new Lockheed U-2 surveillance plane into Soviet air space. Now he gave the okay. From 1956 into 1960 the CIA conducted a number of reconnaissance missions. Among other things, U-2 photographs tipped the CIA to the existence of the new Cosmodrome in 1957. The U-2 had a range of about three thousand miles and a top altitude of around seventy-two thousand feet, which put it, like LeMay's high-altitude bombers, beyond the reach of MiG interceptors. The Soviets helplessly tracked the flights on radar.

Until May 1. Powers, an experienced air force and CIA pilot, took off that day from Pakistan on what was to be the longest-yet U-2 run over Soviet territory. Photographing the Cosmodrome was a priority. Korolev was launching R-7 test flights from there in preparation for a first manned flight, and the CIA wanted to see what could be seen. Powers flew over the site, took his photos, and was heading northward from there when two new Soviet surface-to-air missiles shot him down over the Urals. He managed to eject and parachute to the ground, where he was immediately captured.

When the Americans knew that Powers had gone down, but did not yet know how or where, Eisenhower claimed the U-2 was a NASA weather plane. Khrushchev produced Powers's confession that he was CIA, and angrily demanded an apology. Eisenhower refused.

The timing was most unfortunate. In the first place, May 1 was one of the biggest holidays on the Soviet calendar: May Day, when the whole country was supposed to salute the workers of the world, and all the leaders gathered in Red Square to view a gigantic parade. Flying a U-2 over their heads was an unforgivable insult. Then, two weeks after the incident, Eisenhower, Khrushchev, and representatives of England and France converged in Paris for a summit meeting to discuss increasing tensions in Berlin as well as the potential for a nuclear test ban treaty. Khrushchev scolded Eisenhower in public and then marched out, the summit over as soon as it began. An Eisenhower trip to Moscow planned for June was canceled. Powers's quick show trial in August inevitably ended with him convicted of espionage and sentenced to ten years in prison. He would be returned to the US in 1962 in a swap for the Brooklyn-based Soviet spy William Fisher, aka Rudolf Abel.

Eisenhower forbade U-2s ever to fly over Soviet territory again. The need for new ways to spy on Soviet activities helped spur American satellite development, as well as the air force's top secret Manned Orbiting Laboratory (about which more later).

The UN General Assembly meeting of 1960 began in September. Khrushchev was not invited, but he declared himself the head of the USSR's UN delegation and came anyway. This was a very different Khrushchev from the one who'd charmed the Americans a year earlier. Still fired up, Khrushchev was at his most boorish, argumentative and insulting. The press called him Hurricane Nikita. At the assembly meeting, usually noted for their drowsy decorum, he acted like a bratty child, famously pounding his desk and shouting to disrupt speeches he didn't like, marching down the aisle to wag his finger under the nose of a delegate he disagreed with. He called Eisenhower "a liar" and UN Secretary General Dag Hammarskjöld "a fool," and declared the Security Council "a spittoon." Regarding Powers, he told the press that "the imperialists have poked their snouts in, and we gave them one in the snout."

Then he got his turn to address the assembly. The October 24 *Time* reported:

Ostensibly, his speech was a plea for "complete and immediate" disarmament, but it came out as a threat. His words dropped heavily into the hushed chamber beside the East River: "We will not be bullied, we will not be scared. Our economy is flowering, our technology is on a steep upturn, our working class is united in full solidarity. You want to compete with us in the arms race? We will beat you in that. Production of rockets is now a matter of mass delivery—like sausages that come out of an automatic machine."

Now he was waving his stubby arms. "Of course, you are going to complain all over the place, 'Khrushchev is threatening!' Well, he is not threatening. He is really predicting the future. . . . The arms race will go on, and this will bring about war, and in that war you will lose, and many of those sitting here will not be found any longer—and not many, but perhaps all. You are accustomed to listen to words that lull you. But, as for Khrushchev, I do not wish to pat your heads when the world is on the verge of catastrophe. You want to listen to pleasant words. Well, if these words are unpleasant, that means I have achieved my purpose. That is exactly what I intended."

Once again, he was bluffing—hugely. Rather than factories cranking out ICBM sausages (an interesting metaphor, given how much he'd admired America's capacity for actual sausage-making just a year earlier), he had on hand only four of Korolev's R-7s. *Four.* And he well knew they wouldn't be much use in a shooting match. People in his own country heard Khrushchev's bragging for what it was. A gallows joke made the rounds in Moscow, which had him boasting: "The USA is standing on the edge of an abyss. We are going to overtake the USA!"

Hoping to solve his ICBM shortage, Khrushchev had in 1959 split Soviet rocket development into two components. Korolev continued to lead efforts to put objects and humans into space. The R-7 was fine for that task. But to develop ICBMs, Khrushchev appointed Marshal Mitrofan Nedelin, the stiff-necked war veteran who was almost flattened by that wayward missile at Kapustin Yar back in 1953. As commander of the Strategic Rocket Forces, he went right to work overseeing development of a new, two-stage ICBM called the R-16, designed by Mikhail Yangel, one of Korolev's major competitors. The R-16 could be launched in two or three hours—much faster than the R-7, but still much slower than the US Air Force's Minuteman. "Before we get it ready to launch," one of Khrushchev's generals moaned, "there won't be even a wet spot left of any of us." The arrangement chafed Korolev's prodigious ego, but it also freed him to focus on what he really wanted to do anyway.

The minute Khrushchev got back to Moscow from his tantrums in New York, he summoned Nedelin and Yangel and asked how development of the R-16 was going. He had bragged himself into a corner and really wanted to show the world a working ICBM. Nedelin promised the rocket would have its first test launch before the anniversary of the revolution, November 7. At the Cosmodrome on October 24, Nedelin, under pressure and obstreperous, personally rushed engineers and workers through the fueling and final preparations of an R-16. Where Korolev swore by liquid oxygen and kerosene as the fuel for his rockets, Yangel's R-16 used a more potent and highly toxic mix of nitric acid and unsymmetrical dimethylhydrazine or UDMH, known in the space program as "Devil's venom." Launchpad workers hated being around the stuff and its toxic fumes. The advantage of using it in an ICBM was that the R-16 could stand ready, fully fueled, unlike the R-7.

Yangel, party officials, and a group of military academy students were on hand for the glorious R-16 launch, bringing the number of people at the pad to around 250. According to the documentarian Vladimir Suvorov, Nedelin was such a manly blood-and-guts Red Army squarehead (Suvorov puts it more respectfully) that rather than join every sensible person on the site in the bunker or other safe areas, he insisted on *sitting on a stool practically right under the rocket* when it blasted off. Clearly his earlier near-death experience had taught him no lessons. Military protocol required that all the lower officers and students join him. Yangel, on the other hand, had just strolled away for a smoke when a faulty switch ignited the engines in the second stage of the rocket. An earth-shaking explosion and mushrooming column of burning fuel that was seen from many miles away engulfed the pad and gantry, instantly incinerating Nedelin and those unfortunates with him. Workers falling from the top of the launch tower burst into flames. Others were crushed as the fiery rocket fell over on them, or they choked to death inhaling the poisonous fumes of the Devil's venom. One worker managed to avoid the flames at first, only to get snared in barbed wire that for some paranoid security reason surrounded the launchpad. The fire caught up with him. Black-and-white automatic camera footage, which was not seen in public until after the fall of the Soviet Union, shows workers burning like human torches as they run off the pad. Of the 250 or so people on the site, it's believed that at least 165 were killed, making it by far the largest known accident in the history of space exploration to date. Of Nedelin, "Nothing was left of him, not even a handful of ashes," Sergei Khrushchev wrote. "They found only half of his shoulder strap and the half-melted keys from his office safe."

Chertok wrote that the survivors "were so morally devastated that they had to have emergency psychiatric care." The shocked and demoralized Yangel called Khrushchev to report the disaster. Khrushchev listened quietly, then asked a chilling question: "*And how come you survived?*"

The Soviet government, characteristically, covered up the event. Nedelin and other officials of rank were said to have died in plane crashes. The families of lower-ranking dead heard no official explanations at all. There were rumors outside the Iron Curtain, and a US satellite took photos of what looked like large-scale explosion damage at the Cosmodrome, as well as an unusually large number of aircraft flying in and out—most likely, the CIA would later conclude, "filled with caskets, consultants, and medical personnel." But the Russians would not publicly concede the truth of the disaster until the 1990s, when a memorial was finally erected.

On October 24, 1963, exactly three years after the Nedelin catastrophe, one of Korolev's R-9 missiles would blow up on the launchpad, killing eight workers. Since then, October 24 has been known as the "Black Day" at the Cosmodrome, when no launches or significant work of any kind are carried out.

6

THE COMRADE WHO FELL TO EARTH

The Soviet Union has launched the first man into space. A Ukrainian shepherd, standing on top of a hill, shouts over to another Ukrainian on another hill to tell the news.

"Mykola!"

"Yes!"

"The moskali [derogatory slang for Russians] have flown to the Moon!"

"All of them?"

"No, just one."

"So why are you bothering me?"

On the morning of April 12, 1961, Yuri Gagarin fell out of the sky onto a quilt of farmland growing wheat and rye. It was a collective farm, a *kolkhoz*, called Lenin's Path. They had uplifting patriotic names like that, which they rarely lived up to, making them fodder for surreptitious sarcasm. He unhooked his parachute and strolled, waving, toward a woman and her five-year-old granddaughter who were weeding a potato patch. A calf was tethered near them. All three of them stood still and stared as he approached. Born on a collective himself, Gagarin took care not to frighten them. He took off his helmet to show them his smiling face, but also held it so they could see the big red CCCP blazoned across it. Gagarin called out that he was a friend so they could hear him speaking Russian.

"Have you come from outer space?" the woman asked him.

"As a matter of fact, I have!" he answered with a grin. And then, because his radio had broken and he needed to report in, he asked where the nearest telephone was. She didn't seem to know. The first human being to go into space couldn't report his achievement because he couldn't find a phone.

Gagarin's happy-go-lucky manner belied how haunted by death and fraught with danger his flight was. Just a few minutes earlier he'd been in serious trouble. As it did so often in the Wrong Stuff years, the Soviet space program had just snatched an historic victory from the jaws of catastrophe. And as they also did with regularity, they covered up the fact that Yuri Gagarin, the first man in space, nearly died returning to Earth. Hints and rumors would circulate, but the facts wouldn't be widely known in the West until 1996, when, oddly enough, an auction of Soviet memorabilia at Sotheby's in New York blew the cover.

To put a man in space, Korolev and Khrushchev had to balance their dreams with the needs of Khrushchev's generals. Khrushchev very much wanted another space age propaganda coup, and Korolev was eager to oblige. For him, putting a man into space was a step toward his ultimate goals of putting men on the Moon and Mars. They both knew they were racing the Americans to glory. For Korolev there was the added incentive of racing against time and a failing body broken by Stalin years earlier.

To Khrushchev's generals, shooting a man into space was little more than a publicity stunt, a bigger version of shooting a circus acrobat out of a cannon. They wanted Khrushchev's rocketeers to focus on making more ICBMs, and on spy satellites to gather mapping data of the US to guide those ICBMs to their targets.

In 1959 Korolev cannily offered to build a space vehicle that could do double duty, with a pressurized cabin that could carry either humans or spy cameras and safely return them to the ground. To aim the cameras with the kind of accuracy the generals needed would require gyroscopic stabilizing gear and infrared sensors. It would be a vehicle that could fly itself with some help from ground control, whether it carried a camera or a cosmonaut. With a person in it they called it Vostok; with three people stuffed in it, Voskhod; as a spy satellite it was called Zenit. But it was all the same vehicle.

In their design for the Vostok-Voskhod-Zenit, Korolev's team basically just made a bigger version of the Sputnik sphere. They reasoned that a ball, weighted and heat-shielded on the bottom, should naturally drop correctly. But although they could hurl the ball out of the atmosphere, they couldn't guarantee with any real conviction where or even if it would come back down. Korolev's engineers considered it lucky to hit anywhere within two hundred kilometers of a designated target, and many cosmonauts would find themselves "dusting down" much farther away than that. Starting with Gagarin's flight, official reports would routinely lie about this; wherever the cosmonaut landed, the report would call it the planned location. This led to a running joke among cosmonauts about nailing their "planned unplanned" landings.

The roughly conical shape of the Americans' Mercury capsule meant it would pitch and yaw, so that the astronaut in it would need to do a bit of piloting to get the crucial angle of reentry into the atmosphere correct. Mercury astronauts faced panels of fifty-six switches and seventy-six indicators. The cosmonaut, a mere passenger in his big BB, had just four switches and dials, and they only controlled a tape recorder, internal lights, a small fan, and radio volume. Suvorov once asked Korolev if it wouldn't be frightening to fly in such a new and complex machine, and the Chief Designer tellingly replied oh no, anyone could fly it. No kidding. In an emergency the cosmonaut could execute a few simple maneuvers, but first he'd have to enter a three-digit code, communicated to him by ground control, that unlocked these rudimentary controls. The rationale was that no one knew how being in space would affect a man's mental state. What if he panicked or got the space crazies?

Soviet scientists and engineers argued about how long the first manned flight should be. Some said it would be safest to start with a short, up-and-down suborbital flight, as in fact the Americans would. Korolev disagreed. Why go to the risk and cost of squirting a capsule up and down in fifteen minutes? "The results will be very meagre," he argued. "Science needs a flight around the Earth. Not a small step forward, but a stride forward that is decisive and daring."

Even though the first cosmonauts would mostly be passengers on their missions, all the candidates originally chosen for the program were air force pilots. The thinking was that jet pilots had proven dexterity and excellent vision, and some experience with such spaceflight-like conditions as g-loads and hypoxia, not to mention ejection seats. And of course as fighter jocks they all felt the need for speed, though no one had yet experienced the kind of speed that would come from having giant rockets fired off under your butt. The ones considered for the program were all between the ages of twenty-five and thirty-five. To fit into the tight Vostok space, they could be no taller than five foot six, and had to be trim and light.

With these criteria in hand, recruiters fanned out to air force bases in the fall of 1959. At the Murmansk air base on the Barents Sea, a dozen young MiG pilots were interviewed by a mysterious team of these humorless gray-haired men from Moscow. They were grilled for two days on their medical histories, their backgrounds, and their experiences as pilots. Finally, they were stunned to hear that they were candidates to be the first humans to go to outer space. Four were sent to Moscow for continued screening. Three were rejected. The only one accepted was twenty-six-year-old Lieutenant Yuri Gagarin.

Proud to serve and eager to please, Gagarin was a small young man, five foot three, with bright blue eyes and an ever-ready grin that belied his rough upbringing. He was born in 1934 in an ancient hamlet called Klushino in the Smolensk region. His official autobiography, *Road to the Stars*, penned for him by a state-appointed ghostwriter from *Pravda*, begins:

> I was born into a simple family that differs in no way from millions of other working families in our land of socialism. My parents are plain Russian people for whom the Great October Socialist Revolution had opened up the way to a new and promising life.

It continues in that vein, like the voice-over for one of the singing-peasants propaganda films the CCCP rolled out to hoodwink the world in the 1930s. The truth was much harsher. On the border with Belarus west of Moscow, Smolensk had long been in the path of invaders, traversed by Mongols, Napoleon, the Bolsheviks, and, in Gagarin's boyhood, the Nazis. His parents worked on a collective farm. Drunkenness ran in the family; his father's father had been a notoriously violent alcoholic, and his father downed whole glasses of vodka in a single shot. Yuri grew up to be quite fond of liquor himself.

He was seven when German soldiers occupied Klushino in October 1941. They stayed for three years and commandeered the Gagarins' little home. The family moved into the earth dugout where they'd stored tools, after the Germans confiscated those as well. An older sister and brother were sent to Germany as forced labor, and their father was forced to work for the occupiers. Yuri once watched helplessly as a drunk Gestapo officer hanged five-year-old Boris Gagarin from a tree by his scarf. Their mother got him down in the nick of time, but apparently the psychological trauma never healed.

Boris would struggle with alcoholism and commit suicide in 1977 at the age of forty-one—by hanging himself. Interviewed in the 1970s for science journalist[*] Yaroslav Golovanov's giant, worshipful memorial book called *Our Gagarin*, a tractor driver who grew up with Yuri said, "It was not much of a childhood. There was the war, all the shortages." They pulled apart unexploded bombs to sell as scrap to buy sweets.

Soviet forces bombed Klushino to flinders to drive the Germans out in 1943. The family moved to the nearby city of Gzhatsk. Yuri's interest in flying was piqued when he encountered a Russian fighter pilot who crash-landed nearby. After the war he attended technical schools in Moscow and then in Saratov, on the banks of the Volga. In 1951 he joined a physics club at the school, where he learned about Tsiolkovsky and the Cosmists, and read Verne and Wells. He also joined a local flying club. In 1955 his instructors picked him to be their first student to fly solo—not because he was demonstrating great skill in their Yak-18 trainer, but because he was the most gung ho, as well as the most photogenic for the local Komsomol youth newspaper that covered it. He sent a copy to his mom. She warned him not to get a big head, advice he apparently forgot later. By 1956 he was flying MiGs as an air force cadet. His ingratiating go-getter attitude made him a favorite of his superior officers, but once got him so badly beaten up by his fellow cadets that he spent a week in a hospital bed. Don't look for it in his book. He was a mediocre pilot—he had to sit on an extra cushion to see out—but that wouldn't matter much for a cosmonaut, and his small stature would be a plus. He was

---

[*] Since there were no real news agencies in the USSR, journalists were effectively propagandists who wrote only what was allowed. Fake news is not a modern concept.

commissioned a lieutenant a few days after Sputnik 2 carried Laika into orbit in November 1957, and posted to Murmansk. He and his new wife Valentina endured the miserable six weeks of winter darkness and deep freeze in a tiny room near the base.

By February 1960 an initial field of three thousand cosmonaut candidates was whittled down to a group of only twenty. Korolev called them his "little eagles." Gagarin emerged early as one of the front-runners. His chief competitor was Gherman Titov, who came from a similar background but was of a very different temperament. Two years younger than Gagarin, Titov grew up poor in an isolated, often snowbound Siberian village in the region called the Altai Krai. His father, a schoolteacher, built the family's one-room log cabin. Gherman slept on a shelf above his mother's narrow bed. A sister would later say that maybe it was sleeping up near the ceiling that gave him his first dreams of flying. His father filled the little home with books, and Gherman grew up to be unusually literate for a fighter pilot. He wrote poetry and recited Pushkin at length. An uncle who was a World War I flyer inspired him to join the air force. He earned his wings on his twenty-second birthday. Where Gagarin was a middling pilot, Titov was an ace. Unlike Gagarin, who always looked like his uniform was a little too big for him, Titov looked sharp, natty, well-tailored. And while Gagarin could be friendly as a puppy, Titov could be argumentative to a point that nearly derailed his career more than once when he popped off at superior officers. He and Gagarin admired each other the way opposites do. When it grew clear that they were the stars of the group, they engaged in a fierce competition to be the first human in space.

Gagarin's closest friend in the group was Alexei Leonov. Like the others, he grew up in harsh circumstances. One of nine kids in a desperately poor family in the middle of the frozen Siberian wilderness, he was a toddler when his father was swept up in Stalin's purges. Neighbors took vicious advantage, robbing the family of what little they had—even making three-year-old Alexei strip off his pants—and banishing them as enemies of the people. Taken in by relatives, Leonov slept on the bare floor under a bed until his father was released and rejoined the family. After the war the family moved to the port city of Kaliningrad—the real one, a relatively cosmopolitan place where Alexei began to learn of the world outside of the USSR and read foreign books like those of Theodore Dreiser and Jack London. "I still believed the Soviet Union was the best country in the world," he later wrote. "But I did slowly realize that there were countries where people had a higher standard of living." He first saw Gagarin at the military hospital where cosmonaut candidates were getting thorough physicals. Gagarin looked up from a book, *The Old Man and the Sea*. Hemingway's friendliness toward the Soviets went back to the Spanish Civil War, and his books were as popular in the cosmonaut corps as Verne's.

Boris Volynov was also a friend. Volynov had the distinction—not necessarily a happy one—of being the only Jew among the cosmonauts. He was born in Irkutsk, the one sizable city in the Jewish Autonomous Region, created by the rabidly anti-Semitic Stalin to get Jews as far away from Moscow as possible—five thousand miles away, on the border with northern China. It's a desolate region where winters are bitterly cold, summers unbearably hot and steamy, and Jews weren't much more welcome than elsewhere in the USSR. Whether anti-Semitism played a role or not, Volynov would have to wait until 1969 to get his first mission—and that one, in the grand Wrong Stuff tradition, would nearly kill him.

Cosmonaut training was run by Colonel Yevgeny Karpov, a specialist in flight medicine, and the controversial air force and KGB general Nikolai Kamanin. Kamanin was one of the most famous people in the USSR. In 1934 he and a few other pilots flew into an uncharted Arctic region to rescue the crew of the steamship *Chelyuskin* after it sank and left them stranded on the ice. The crew hand-built landing strips in the snow so the flyers could save them. These pilots were so highly regarded that a whole new honor was created for them: Hero of the Soviet Union, which remained the highest honor in the land until the end of the USSR. To some in the space program, however, the ramrod-straight and high-domed Kamanin was a humorless old warhorse. Golovanov later described Kamanin as "a malevolent person, a complete Stalinist bastard." He showed his KGB side by constantly scrutinizing the candidates with a cold eye for "ideological reliability," and chaperoning the young men in public. Everything they hated about their training they chalked up to him. Yet as a flyer himself, Kamanin argued constantly with Korolev and the other program managers, unsuccessfully for the most part, against "over-automating" their spacecraft and treating cosmonauts like dogsmonauts. His secret diaries of his time in the space program from 1960 to 1971, first published in 1995, are another key insider's account.

To build the top secret Cosmonaut Training Center—the Tsentr Podgotovka Kosmonavtov, or TsPK—Karpov and Kamanin cleared a large area in the middle of a dense birch and pine forest forty-one kilometers northeast of Moscow's city center, not far from both Korolev's "Kaliningrad" site and the Chkalovsky military air base. An unmarked road ran along the solid wall of trees at the edge of the forest. If you blinked you could miss the narrow gap and small gatehouse at the entrance. A tarmac lane led from there to the hidden compound in the heart of the woods. At first it was just some low wooden barracks and training structures, but over the next decade it grew into a small town, called Green City and then Zvezdny Gorodok, Star City. Early on the cosmonauts and their families roughed it—nothing new for Soviet citizens. The Titovs shared a two-room apartment with another cosmonaut and his wife. Volynov wrote that when he and his wife entered their assigned apartment, "We didn't have one piece of furniture, only a few carpets that were lying rolled up in the corner of the room." That was better than what Leonov and his wife Svetlana got, "bunk beds in the corner of a volleyball court. We had to drape newspapers over the net in order to get some privacy, because another pilot and his wife were sleeping at the other end of the court."

Some of their training was similar to what the Mercury astronauts were going through. They endured high g's in a centrifuge, which once spun out of control, nearly killing a trainee. He dropped out of the program. They experienced momentary weightlessness in parabolic aircraft flights, and practiced in a mock-up of the capsule (not that there was much to practice). Since it was quite possible that on reentry they might come down way off target, they did wilderness training; dropped into an isolated area of forest or mountains, they had to make their own way back to civilization. Gagarin would take off carrying a pistol and a hunting knife to ward off wild bears should he land in a forest, sharks if he dropped in the Pacific, or hostile natives if he turned up in a foreign land. In that case, if interrogated he was instructed to give only his name and say his address was "Moscow, Cosmos."

Then there was the intensive parachute training. Gagarin had only made five jumps before. They jumped from numerous types of aircraft, at various altitudes, in all kinds of weather. They were ejected from fighter jets. "We jumped into trees, on hillsides, into lakes and rivers," Titov recalled. "We fell into marshes and onto hard ground. We landed in wheat fields and on concrete surfaces." One of Titov's jumps was almost his last. He got tangled in his parachute's lines and spun so wildly he nearly lost consciousness, barely straightening out for a very hard landing.

At the time there was simply no alternative to ejecting and parachuting to the ground. A few weeks before the first flight, Kamanin wrote in his diary:

1961 March 21—Spring at Tyuratam
It is a beautiful day. The cosmonauts discuss contingencies in case of a water landing. In fact their chances are slim. There are only two Soviet ships equipped with HF and UHF direction-finding equipment that could locate them. The NAZ [emergency] ejection seat is not designed to float, and the spherical re-entry capsule is no better. Therefore the only option is a landing on the territory of the Soviet Union. . . .

One of Titov's least favorite tests was in the "vibration seat," "a devilish affair that almost drove a man to distraction." It was just the front bench of an old bus, but attached to a device "that produced the wildest vibrations I have ever known. Under full throttle the engine whipped the seat into a paroxysm of vibration that rattled a man's eyes in their sockets. After several seconds the seat was almost unbearable, and even breathing became unbearable." The trainees hated it and coined a slogan, "To the Cosmos—by Bus!"

Much as they disliked some of the other tests, the trainees most hated and feared the isolation chamber, aka the Chamber of Silence and the Chamber of Horrors. Mercury candidates also hated theirs. Titov called it "something out of the worst medieval times." It was a soundproof box mounted on shock absorbers in the middle of a laboratory. The walls were sixteen inches thick. It was furnished with a replica of the Vostok seat, a small bed and table, and an electric hot plate for heating up food. When the door closed you were plunged into total silence. The point was to test trainees' ability to withstand the complete seclusion of a long spaceflight—say, to the Moon and back. It was an exercise in harrowing loneliness, sealed inside for up to *fifteen days*, knowing they were under 24/7 scrutiny. For long periods they endured silence so total, so profound, that their heartbeats boomed like cannons. Then suddenly lights flashed and music blared and they were supposed to solve complex math problems while an amplified voice thundered the wrong answers at them. But worst of all were the oxygen deprivation tests, when the air supply was gradually pumped out while the trainee wrote his name on a pad, over and over. The writing devolved into a scrawl of gibberish as his oxygen-starved brain shut down. He dropped the pen, his head lolled, and eventually he blacked out.

Two years earlier, in the very first episode of *The Twilight Zone*, Rod Serling had tried to imagine the deleterious effects a prolonged stay in such an isolation chamber might have on the mental state of a future spaceman. Cosmonauts used various stratagems to stay sane through the ordeal. Yuri Gagarin survived by keeping up a positive attitude and appearance, cheerfully holding one-sided conversations with his silent observers, singing little ditties he made up about objects in there with him—the hotplate, the squeeze tubes of cheese, even the electrodes monitoring him. They were allowed to bring books. Titov read and recited Pushkin, which struck the observers as a bit highfalutin for a Soviet hero, perhaps even a whiff fruity.

Mercury astronauts were in training at the same time. Leonov later described some of the similarities and differences between the cosmonauts and the astronauts. There were height and weight restrictions on astronauts, though they could be bigger than cosmonauts, up to five foot eleven and 180 pounds. All the original cosmonauts and astronauts were military pilots—except, interestingly, Neil Armstrong. Though Armstrong had flown seventy-eight combat missions for the navy during the Korean War, he had resigned his commission and was now a civilian test pilot. Buzz Aldrin believed that Armstrong was chosen to step down onto the lunar surface first to continue propagating the image that the conquest of space, as far as the US was concerned, was a non-military endeavor. NASA dismissed this claim, but then NASA would.

Another difference was a very big one. Everything about the Soviet program was top secret. Only after a cosmonaut flew a successful mission would his name be made public. The Mercury 7, to the contrary, were selected partly for their suitability as All-American celebrities. They were, of course, all white males, no surprise in the early 1960s. They were older than the cosmonauts, thirty-two to thirty-seven at the time of their selection. They were all married with no divorces, sober, mature, clean-cut, Christian, college-educated, professional-class patriots. And they were promoted as heroes before a single one of them ever flew.

Leonov ruefully noted one other difference. *Life* paid the Mercury 7 half a million dollars for their life stories. A cosmonaut made the equivalent of about one hundred dollars a month after his Communist Party dues were deducted.

Meanwhile, Korolev was driving himself and his team beyond exhaustion as they raced to build and test the Vostok-Zenit-Etc., as well as the modified R-7 needed to boost it into orbit. At one frantic moment one of the engineers, lacking an antenna for the exterior of the capsule, unrolled a metal tape measure and stuck that on the vehicle as a substitute. In May 1960 Korolev's team got an unmanned Vostok into orbit but couldn't bring it back down when its retro-rockets misfired. They launched a second test Vostok that August, with two small dogs aboard, Belka (Squirrel) and Strelka (Little Arrow), harnessed into a cabin that sat on the ejector seat rails. They had company: a rabbit, two lab rats, forty-two mice, and some fruit flies. Also small flaps of skin donated by staff at the Moscow Institute of Experimental Biology, thought to be the first human tissue in space. After four weightless orbits Belka spit up, making her the first Earth creature known to experience space sickness. Korolev was concerned that prolonged weightlessness might make a cosmonaut sick too. Originally he'd been thinking that the first cosmonaut in space should spend twenty-four hours completing sixteen orbits. Now he decided the first mission should just be one orbit.

Otherwise things went smoothly during the eighteen-orbit, day-long flight. This time the retro-rockets worked. The small cabin holding all the animals ejected from the larger capsule on reentry, and both parachuted to the ground. The cabin landed in a meadow. Farm workers wandered over to scratch their heads at it. A small metal tag said *Please inform the Soviet space center immediately on discovery*, but before they could go find a phone the retrieval team's cars were already bucking across the meadow toward them. The cabin had landed less than ten kilometers from its target, a remarkable achievement. Belka and Strelka and all their companions were fine. They were the first Earth creatures to orbit the planet and return safely.

Based on just this one successful test, Korolev decided they could move forward with confidence to a manned flight, which he scheduled for December. Then Nedelin's R-16 disaster that October caused a crisis of confidence in the Soviet space program, much as the horrific Apollo 1 accident in 1963—when an electric spark turned the oxygen-rich environment of their capsule into a horrific inferno that burned Gus Grissom and two other astronauts to death—would do at NASA. But whereas NASA would put manned Apollo missions on hold for almost two years of extensive internal reviews and Congressional hearings, Khrushchev's and Korolev's cravings to keep beating the Americans in space meant a much shorter pause at the Cosmodrome. The next R-16 test was conducted in February, and with no Nedelin around to rush everybody, it was successful. By 1965 almost two hundred R-16s were deployed, and they remained a mainstay of the Soviet nuclear arsenal into the 1970s.

Korolev used the few months' delay to continue the tests of his rocket and capsule. They did not inspire hope. The dogs Pchelka and Mushka were sent up on December 1. The capsule successfully made seventeen orbits in twenty-four hours, but its reentry rocket malfunctioned, meaning the capsule would land way off course, even outside the USSR. A self-destruct system had been built in for just such a contingency, to prevent foreigners from studying the craft. The mission control crew triggered it on reentry, blowing the capsule, and Pchelka and Mushka, to fragments. The official announcement was that the craft had burned up when it reentered the atmosphere at too steep an angle.

Two more dogsmonauts, the huskies Shutka and Kometa, were launched on December 22. When that flight was aborted because of problems during the ascent, the Vostok capsule separated and parachuted down in the frozen Siberian taiga near the Tunguska River, the site where the famous meteorite exploded in 1908. A rescue team set out from the Cosmodrome, while a bomb disposal crew was sent from Leningrad. They had sixty hours before the self-destruct system automatically detonated. Trudging through waist-deep snow and temperatures of 40 degrees below zero, they didn't get to the capsule by the sixty-hour deadline. But in a twist somehow typical of the Soviet space program at the time, the self-destruct system had failed. So had the ejection apparatus, which should have fired Shutka and Kometa out of the sphere. The rescuers opened the cabin to find them barking and wagging their tails, happy to see humans, two of the luckier early travelers in Soviet space.

By this stage Korolev was feeling such pressure to produce a new victory that he suffered his first heart attack. While he was in the hospital they found that his kidneys were malfunctioning as well. Both were common ailments among survivors of Stalin's Gulag camps. The doctors wanted him to take a long rest, but he went back to work. He triumphed on March 9, when a Vostok successfully completed one orbit of Earth and landed at the planned retrieval site. Another husky, Chernushka ("Blackie"), made the trip. Alexei Leonov wrote that Chernushka was a last-minute replacement for a dog that had been "trained to push buttons with its muzzle while in flight to obtain food and water," and then, proving its intelligence beyond doubt, ran away. Chernushka was a stray they grabbed at the Cosmodrome. She handled spaceflight well. When they released her from her harness she ran around in the snow happily barking. The engineers had to chase her for her post-flight medical check.

Nearby, the cosmonaut Ivan Ivanovich, Russian for John Doe, lay sprawled in newly fallen snow. Ivanovich was a dummy cosmonaut. His synthetic leather skin was stretched over a metal armature, his head snapped on, and his face was rubbery and dead-eyed. He wore a cosmonaut's orange space suit, white helmet, gloves and boots. Ivan's job this trip was to test the cosmonaut's ejector seat as the capsule descended. It worked. He made a soft landing by parachute. Vladimir Suvorov was there to get him on camera. "'Ivan Ivanovich' is not injured," he wrote in his diary, "but it feels creepy to look at him lying motionless in the snow with that fixed glance of those fake eyes and a deadly mask of a face. He looked precisely as if a real cosmonaut had been killed during the landing."

Ivan flew one more successful mission on March 25, with another dog. During the flight the capsule broadcast a recorded message to test the communications equipment. Korolev's team had debated what message to send, given that Western listening posts would surely be dialing in. A voice reciting technical details might be taken for a spy mission. A tape of someone singing was rejected because the eavesdroppers would conclude the cosmonaut had gone space daffy. In the end they went with choral music and a man reciting a recipe for borscht. Rumors flashed around the West of a secret Soviet flight with a rather odd cosmonaut aboard.

Again, Ivan ejected from the capsule and parachuted to the snow. Suvorov writes that peasants from a nearby village came rushing over, thinking he might be another downed U-2 pilot. They flipped open his helmet's visor to see a card that had been placed inside with the word *MAKET* (DUMMY) on it. That didn't stop them from punching the hated American spy-pilot in his rubbery face. Soldiers of the retrieval crew showed up and rescued him.

The record of tests for the Vostok capsule and its R-7 booster was now split nearly evenly: eight of fifteen R-7 launches successful, but only three of eight Vostok flights. The first cosmonaut would have less than a fifty-fifty chance of surviving. Chertok would later say it plainly: *"We had no grounds for optimism."* But Korolev was in a desperate rush. He was very aware that NASA was also training its Mercury 7 astronauts, and testing its rockets and Mercury capsule. According to Vladimirov, he had a special office at OKB-1 staffed with people whose only job was to monitor world press for mentions of NASA and feed him daily reports. He knew it was critical to his high status with Khrushchev to keep one-upping the capitalists. Luckily for him, their record at that point was just as dismal as his, and unlike his, their failures were all mortifyingly public. In July 1960, an Atlas booster failed and was blown up one minute into its ascent. A test the following November became infamous as "the Four-Inch Flight." That's as high as a Redstone rocket got before settling back down on its fins. The escape mechanisms at the top of the capsule popped off, the only part of the system that went airborne. It was called "the most distressing, not to say embarrassing, failure so far in Project Mercury."

The outlook for Mercury improved in January 1961. A chimpanzee named Ham, one of six "astrochimps" trained for space, shot up 157 miles and splashed down 16.5 minutes after taking off. Ham was unharmed, even nonchalant. Still, there had been enough nagging problems during the very short flight that NASA wanted more tests before launching a human.

It showed a crucial difference between NASA and the Soviets. It's very unlikely that Korolev, driven as much by himself as by his political bosses, would have paused for more testing at that moment. In similar situations in the future he certainly did not. NASA had its share of accidents and made its mistakes. That's in the nature of engineering new technologies. Korolev, on the other hand, was often downright reckless, in ways the Americans never would have been.

Based on faulty intel from Soviet spies in the US, Korolev was now convinced the Americans would launch their first astronaut at any minute. In his panic to beat them to it, he decided that the prize was worth risking his young eagles' lives. And on March 23, the youngest of them paid the price.

That day, twenty-four-year-old Valentin Bondarenko was preparing to exit the isolation chamber at TsPK after ten days of being sealed inside. He was Ukrainian, from Kharkov, well-liked by the others, athletic, with a fine singing voice. He just had to wait thirty more minutes as the technicians outside gradually raised the pressure from a high altitude simulation to normal. This had to be done slowly, or Bondarenko, like a deep sea diver rising too quickly, would get the bends. While he waited, the young cosmonaut pulled off his electrodes. He daubed the itchy spots with rubbing alcohol on cotton balls. He accidentally tossed one of the cotton balls onto the hotplate. It burst into flame—which then exploded through the whole oxygen-rich interior of the chamber (another precursor of the Apollo 1 disaster). Bondarenko's screams could be heard outside as his hair caught on fire and his woolen training suit melted into his skin. Horribly, the technicians still had to wait for the pressure to equalize before they could get the door open. Finally they dragged him out, his charred skin smoldering, burned everywhere but the soles of his feet. He was rushed to the nearest hospital, where he broke everyone's hearts as he pitiably repeated, "I'm so sorry. No one is to blame but me." He died eight hours later.

His death, of course, was not made public. But it weighed heavily on the hearts and minds of everyone in the cosmonaut program. And the gloom intensified when, in the final weeks of the training, Gherman and Tamara Titov celebrated the arrival of their firstborn child, Igor, only to discover that he was born with a damaged heart. Igor died, with the cruelest timing, just before the star trainees Titov and Gagarin were told which of them would be the first in space.

On April 8, Kamanin called Gagarin and Titov into his office and matter-of-factly informed them that Gagarin had been chosen to go first, with Titov his backup. "Titov's disappointment was quite obvious," Kamanin wrote in his diary. Titov was the better pilot, but Vostok didn't need a pilot. Dogs, a manikin, and flaps of skin had successfully flown in it. The program managers chose Gagarin because he made a better Soviet everyman than Titov. Khrushchev agreed. He liked that Gagarin had come from a background similar to his own, and was said to distrust Titov's "German" name. "Yura [familiar for Yuri] turned out to be the man that everyone loved," Titov said years later. "Me, they couldn't love. I'm not lovable."

On the night of April 11 Gagarin and Titov bunked in one of the cottages. Since Gagarin's flight the next day it has been preserved exactly as he left it, a nearly sacred pilgrimage site that all crews visit before they take off. For dinner the two cosmonauts squeezed tubes of space food—pureed meat and chocolate—into their mouths. They shot some pool, Titov recited some poetry, and they lay down on side-by-side beds around ten o'clock. When Yevgeny Karpov checked on them a little later they both seemed in deep sleep, but they'd confess later they didn't sleep a wink.

Neither did Korolev. In April 2011, the Russian space agency Roscosmos would release some previously secret documents for the fiftieth anniversary of Gagarin's flight. One thing they revealed for the first time was that right up to the day before Gagarin's flight Korolev and his team were worried that the modified R-7 didn't have quite enough oomph to loft the Vostok capsule into orbit. That day Korolev had technicians pull some gear out of the capsule to shed thirty pounds. In the process they caused an electrical short, and spent the whole night tracking it down and fixing it.

Karpov roused Gagarin and Titov at 5:30 a.m. Then Korolev looked in on them. He was showing the strain. "It was the first time I had seen him looking care-worn and tired," Gagarin would note. "He had clearly had a sleepless night. I wanted to give him a hug as if he were my father." After more space food they got into their layered gear: long underwear, then pressure suits, then orange spacesuits and white helmets with CCCP painted on them in red. The CCCP was a last-minute addition, hand-painted above the visor by a launch technician so that when Gagarin landed out in the boonies the locals wouldn't mistake him for another American spy and attack him the way they had Ivan. In these cumbersome outfits they climbed up into a bus at 5:45 for the ride to the launchpad. Kamanin rode with them.

One of the best-known anecdotes from *The Right Stuff* tells how astronaut Alan Shepard, stuck in his Mercury capsule on the launchpad through hours of delays, couldn't hold his pee any longer and went in his flight suit. Less familiar outside Russia is a similar tale from the Cosmodrome about how Yuri Gagarin dealt with his own need, a week before Shepard. On the way to the pad Gagarin asked the driver to stop the bus. He clambered out, walked to the right rear wheel, unzipped his pressure suit, and pissed on the tire. One explanation is that he knew that once he was strapped into Vostok it would be several hours before he'd get another chance to pee. Another is that he was following a superstitious ritual begun by Russian fighter pilots during the Great Patriotic War. Cosmonauts have their own superstitions. For decades, it was a rite at the Cosmodrome that all cosmonauts must get off the bus and pee on the right rear tire before a flight—they had to boldly go, at least one journalist joked. Female cosmonauts poured their pee out of a cup. In 2019 it made international news when Roscosmos decided to redesign its standard flight suit without a crotch zipper.

Gagarin said goodbye to all his friends, rode up in the elevator, and climbed into the small capsule. The engineers closed the hatch behind him. Then in fact he did sit for two hours as one lingering glitch after another was rectified. At one point a warning light showed that the hatch hadn't properly sealed. The engineers unscrewed thirty-two bolts and opened it. Gagarin looked out at their worried faces, every one of them with a cigarette dangling from their lips, and quipped, "What, you need a light?" For all his habitual cheerfulness, Gagarin knew his chances were fifty-fifty at best. Two days earlier, he'd written a letter to his wife, to be opened only if he died. "If something should happen, I beg you, Valiusha, do not die from grief. Things happen in life and there is no guarantee that tomorrow you won't be run over by a car."

At 9:07 a.m. Gagarin felt the engines kicking in and called out, "*Poyekhali!*" ("Let's go!") In the recording he whoops it like a kid at the start of a roller coaster ride. To this day, "*Poyekhali!*" is a popular toast with Russian vodka drinkers. Korolev's temperamental R-7 lifted off smoothly into a warm, cloudless blue sky. Everyone in the control bunker started to breathe again. But the first glitch came soon. The rockets didn't cut off when they should have, and shot Gagarin up to an altitude of 203 miles instead of the planned 143. Soon though he settled into his single 108-minute orbit. Down below, Korolev's hands trembled and he was so ashen the whole flight that the team worried as much for him as for Gagarin. When the Kremlin was informed Gagarin hadn't blown up (yet), they awarded him a field promotion—space promotion?—from lieutenant up to major. State radio was then allowed to broadcast the triumphant flight of Major Yuri Gagarin. Everyone everywhere in the Soviet Union heard the announcement at the same time. A direct inspiration to Orwell, in the Soviet Union there was one radio station, All-Union Radio, "and it played constantly into every nook and cranny of public and private space," historian Andrew Jenks explains. "Soviet apartments were wired for radios *that could not be turned off*. [Emphasis added.] Radios blared constantly in courtyards, school yards, factory floors, and street corners," as well as out on the collective farms. Television was far less common, with five million sets in a country of around two hundred million. They cost about two months' salary and were so poorly made that the Interior Ministry issued instructions on how to prevent your TV from overheating and blowing up.

At seventy-nine minutes into its flight, Vostok was arcing smoothly over West Africa, its passenger musing about how his dear old mum was going to react when she heard about this.

Cosmonauts were not allowed to tell even their families of upcoming missions, so any deadly failure could be covered up. At this point the engineers on the ground triggered the retro-rockets to brake for reentry. The rockets burned as planned—then things went wrong again. "As soon as the braking rocket shut off, there was a sharp jolt, and the craft began to rotate around its axis at a very high velocity," Gagarin would tell Korolev and Kamanin the next day. "Everything was spinning around."

Behind the Vostok sphere that held Gagarin was an equipment module with the braking rockets, oxygen tanks, and batteries. This was supposed to detach when the retro-rockets cut off. It had not. A thick cord of electrical cables kept them "tied together, like a pair of boots with their laces inadvertently knotted," Jamie Doran and Piers Bizony write in *Starman*, their biography of Gagarin. "The whole ensemble tumbled end over end in its headlong rush to earth." If the two modules collided as they lurched around, Gagarin would likely be killed. Vostok was weighted so that the thick heat shielding would face downward during reentry. But the attached unit kept the capsule from orienting properly. Vostok was a plummeting fireball pitching and yawing uncontrollably. Gagarin could feel the intense heat and hear the hull crackling. His vision blurred as heavy g-forces slammed him around.

Then came a bit of luck. The cables burned through, and the capsule broke away from the rocket pack. Unfortunately, that made it start spinning so violently that Gagarin nearly blacked out. "The indicators on the instrument panels went fuzzy, and everything seemed to go gray," he'd report. As the capsule dropped through denser air the fire burned out and the spinning eased somewhat. Gagarin could see blue sky out the charred porthole. The ejection device was supposed to be triggered automatically at seven kilometers, but the indications are that Gagarin decided not to wait. Apparently he blew the hatch manually and ejected early. There was a rumor that he'd panicked. But maybe in the midst of being tossed around in a superheated metal ball he made the logical decision not to trust the faulty hatch and not to bet his life that the automatic ejection device would function properly.

Khrushchev was vacationing at his opulent dacha in the Black Sea resort of Pitsunda. He was supposed to be relaxing, but he was panicky when he got Korolev on the phone. "Is he alive? Is he sending signals? Is he alive? Is he alive?" Korolev was so relieved, and maybe surprised, that Gagarin had survived that he shouted at Comrade Khrushchev at the top of his raw voice, "HE'S ALIVE! HE'S ALIVE!"

Gagarin never mentioned to the press or in the relentlessly rah-rah *Road to the Stars* the technical failures that could have killed him, just as he dutifully kept mum about the whole business of ejecting from the capsule. But when Korolev and Kamanin debriefed him the next day he described in detail what happened. The transcript of that meeting was locked away. It wasn't until 1991 that a Moscow newspaper published it. There was little notice in the West until a Sotheby's auction of Soviet space memorabilia in 1996.* One of the hundreds of items was a set of notes Karpov had furtively scrawled during the flight:

> *Malfunction!!!*
> *Sudden impact.*
> *Don't panic!*
> *Emergency situation.*

It was a good thing for Karpov that the KGB never found out he'd made these notes.

Gagarin and Vostok came down two kilometers from each other, and about five hundred kilometers off course. As he drifted down under his parachute, Gagarin had no idea how lucky he was that his parachute opened: later it would come to light that the engineer in charge of testing cosmonaut parachutes failed to either report or fix a problem with them snagging on an antenna as they deployed. With Korolev rushing his team so, this person decided to keep his mouth shut and hope for the best. Unaware that he'd dodged yet another bullet, Gagarin gazed below him and was amazed to recognize the terrain. Below him curled the silver ribbon of the Volga, and on opposite banks lay the twin cities of Engels and Saratov, where he'd had his first flying lessons as a youth back in the early 1950s. What were the odds? He landed in a softly plowed field near a tiny ramshackle village called, appropriately, Audacity (Smelovka).

---

* There was another Sotheby's connection to the Soviet space program. In 1994, Ivan Ivanovich was sold at auction there.

Children from the village saw the Vostok ball hit the ground, bounce a little, roll a little, and come to rest on its side near the river. Scorched black from the heat of reentry, its open hatch gaping, it didn't look like an historic victory. It looked like an old and battered object raked out of a tragic fire and then discarded. Schoolboys clambered inside and filched tubes of the cosmonaut space food, which they passed around. Adults joined in, pulling anything they could from the capsule—radio antennae, parachute silk. A squad of military police arrived and tried to ward the people away with threats that the Vostok might explode. They threw a tarp over it and a cordon around it before the villagers could roll the whole thing away. Doran and Bizony write that "a large contingent of KGB officers" later descended on Smelovka and went house to house sternly ordering the return of everything the villagers had snagged. This included an inflatable life raft a local fisherman had stolen to use in the river. He gave it back, complaining with exquisite peasant logic that anyway it had holes in it and was not much use to him.

With the retrieval team so far away, an officer from a nearby air base was hustled over with a contingent of soldiers to collect Gagarin. One soldier handed Gagarin his Communist Party membership card and a pen and asked Gagarin to sign it—his first of many, many autograph requests. A giddy, ebullient Khrushchev spoke to him on the telephone, telling him he was being awarded a Hero of the Soviet Union star and crowing, "Let the capitalist countries catch up with us!"

The next day, three helicopters landed in the field near the Vostok capsule. Korolev himself emerged from one, grinning broadly. The capsule was hooked up to a helicopter to be flown away. Some of the retrieval crew decided to mark the historic spot. They pounded a crowbar into the earth and etched *12.VI.61* onto it. One of them produced a bottle of Stolichnaya and they all toasted the first comrade who returned from outer space.

Their ad hoc monument was quickly removed. The Soviet government wanted an official world record from the Fédération Aéronautique Internationale, the French organization that was in effect the Guinness for aeronautics. It meant telling a few lies. They had to lie that Gagarin landed in his capsule, not separate from it, because that was an FAI rule for record flights. The Soviet Union did not want it known that it could shoot a man into space but had not figured out how to land him in his capsule. Gagarin would hem, haw, and dodge questions about it from suspicious foreign journalists. They also lied that he landed on target, not five hundred kilometers away from it.

But the biggest lie was about where he'd taken off from. The Soviets weren't aware that the CIA had known since 1957 exactly where the Tyuratam Cosmodrome was. In a pitiably vain attempt to keep its location secret, they said Gagarin had taken off from a launch facility at Baikonur, a mining town that was in fact some three hundred kilometers north of Tyuratam. From then on the Tyuratam Cosmodrome has been known as the Baikonur Cosmodrome, even though it hasn't moved an inch closer to Baikonur in all these years. They continued the charade even after the Cosmodrome was open to visits by foreign press and dignitaries in the 1970s. And, to disguise how much their space program was a military venture, the military personnel dressed as civilians during the visits.

Gagarin went into the history books as the first human to orbit the Earth, but in fact where he dusted down was a bit short of a full orbit. The real first human to fully orbit the planet would be Gherman Titov, but he would spend his life relegated to being Gagarin's also-ran.

It was just past 1:00 a.m. EST when US radar stations spotted a launch at the Cosmodrome. By 1:15 radio monitors were eavesdropping on a land-to-space conversation between "Dawn" (Korolev's code name) and "Cedar" (Gagarin). Their identities were unknown, but it was unmistakably a communication between Soviet ground control and a man in space.

Moscow announced that Gagarin had safely landed at 5:30 a.m. EST. In a 1962 profile of NASA spokesman John "Shorty" Powers, *Time* reported that "John G. Warner, a young United Press International rewrite-man in Washington," roused Powers and peppered him with demands that he wake one of America's Mercury astronauts for comment. "At last, his patience exhausted, Shorty blurted the ill-advised statement that no one has since let him forget: 'If you want anything from us, you jerk, the answer is that we're all asleep.'"

Warner got his own back with the headline:

SOVIETS PUT MAN IN SPACE
SPOKESMAN SAYS U.S. ASLEEP

On April 14 a four-propeller Ilyushin Il-18 airliner, with an escort of fighter jets, circled and set down at Moscow's Vnukovo Airport. Major Gagarin came down the steps and then began a long march across a red carpet laid on the tarmac from the plane to a reviewing stand. A beaming Khrushchev awaited him in a crowd of officials, as well as Gagarin's family. In the newsreel footage the newly minted major marches smartly, looking a bit like a toy soldier with his boyish face and small stature. He's wearing a new officer's overcoat that's tailored with military precision and yet still looks a little too big for him, like he's a kid playing army in his dad's coat. As he made his interminable way across the very long red carpet, a different wardrobe malfunction made some in the crowd, including the filmmaker Suvorov and the journalist Evgeny Riabchikov, gasp and stare. The long laces on one of his new boots had come untied and were flapping behind him. The tension that sight produced—that the first human to soar around the world might trip over his own bootlaces—makes it another quintessential Wrong Stuff moment. Riabchikov recalled what torture it was to watch. Gagarin was aware of it too, and decided to keep going and hope for the best. In its way it sums up the Soviet space program, a small echo of the "bootlaces" of cables that had tied his capsule to his equipment pod and almost killed him. Suvorov recorded in his diary that there was much discussion later about editing the bootlaces out of the footage. The official film of the event would in fact be edited quite drastically a few years later, but the flopping bootlaces stayed in.

On the platform there were heaps of flowers, and everyone was dressed up for the occasion, in blocky, fashion-damaged Soviet style. Except for Gagarin's mother and father, who looked like the Soviet version of Grant Wood's *American Gothic*, a pair of dour farm types who didn't seem at all pleased or comfortable to be sharing the platform with the leaders of the Soviet empire. Khrushchev had specifically ordered them not to dress up. He wanted it visually clear that the hero Gagarin came from humble peasant stock, like he did. The

April 21 issue of *Life* ran a photo on the cover of Khrushchev hugging Gagarin, and nineteen full pages of reportage inside, amid ads for Sanka and NoDoz, Old Grand-Dad and new Fords. Khrushchev sent his own sartorial signal by wearing a battered porkpie hat that looked like he'd stepped on it before cramming it on his head. It was too small for his bulb of a skull, a silly hat, a clown's hat. We can presume he chose it with intent, a way of playing himself down, just another Soviet Joe in a crumpled old hat, at the same time he was triumphing over the whole world. It was an iteration of the aw-shucks bluff that had gotten him through the deadly Stalin years and was so confounding to the West now.

In the newsreel footage, Chief Designer Korolev can be spotted, if you know what to look for, at the edge of the gaggle of officials, in a dark coat and fedora, watching the ceremony like any other nobody. Although he had vastly more to do with Gagarin's historic flight than Gagarin did, he got no hugs or kisses on the cheek, not in public. After a short ceremony at the airport a long motorcade headed to the Kremlin. The Khrushchevs and Gagarins rode in a convertible Zil with the top down, like the Kennedys in Dallas. The boxy Zil was a knockoff of a mid-1950s Cadillac. The Chief Designer trailed the procession in a dowdy old Chaika. Chaikas made Zils look glam. The fan belt snapped on the way to the Kremlin. Crowds from the villages along the way strew the asphalt with flowers. As a sign of how crazy Gagarin's celebrity was about to get, someone also threw his bicycle in front of the Zil, trying to slow it down so he could catch a photo.

The enormous crowd that awaited them in Red Square and spilled out to the surrounding streets astounded everyone. It was said to be the largest spontaneous celebration in Moscow since the end of the Great Patriotic War. "Nothing like it had ever happened that I could remember," Sergei Khrushchev later said. "As far back as Stalin's day the KGB was panic-stricken at the thought of a crowd." When there were public demonstrations, they were strictly choreographed, with cold-eyed KGB and armed soldiers vigilant for the slightest deviation from the program. Unplanned street parties were unthinkable. Jokes like the one about the Ukrainian shepherds suggested that not everyone in all the Soviet republics was quite so impressed with Gagarin's achievement. But what looked like a million or more Russians jammed Red Square, waving up at Gagarin and Khrushchev and the others standing on top of Lenin's mausoleum, grinning and waving back at them. Soviet citizens didn't just rejoice, they gloated. A letter-writer to one newspaper crowed, "Hey Uncle Sam, listen to the rocket thunder over the entire world and over the territory of California and your various Floridas."

Gagarin gave a patriotic speech, frequently interrupted by a roar of cheers from the crowd, winding up with, "Glory to the Communist Party of the Soviet Union led by Nikita Sergeievich Khrushchev!" Khrushchev kvelled, wiping joyful tears. The space program—truly, *his* space program—had given him yet another history-making triumph. Let his enemies in the West, as well as doubters in the worldwide brotherhood of Communism (like China's Chairman Mao, who tended to treat him like a buffoon and a putz), behold and be awed. Afterward there was a reception at which the champagne and vodka flowed like rivers and pretty much everyone got drunk, with more than a few needing to be carried out as Khrushchev's toasts wobbled into incoherence.

★ ★ ★

The whole world cheered the Soviets' achievement. The whole world, except the United States. For NASA, Gagarin's triumph was even more demoralizing than Sputnik had been. They were just weeks away from putting the first Mercury astronaut in space, and once again the Soviets had trumped them. Had they shown just a little more of the Wrong Stuff after Ham the chimp's flight, been a little more like the Soviets in their risk-taking, they might well have put a man in space before Gagarin got there.

Then the other shoe dropped. The US-backed invasion of the Bay of Pigs in Cuba began on April 17, just five days after Gagarin's triumph, and was over by April 18. When Fidel Castro and his revolutionaries had taken over Cuba in 1959 and started edging toward a friendship with the Soviets, Eisenhower and the CIA began plotting how to oust him. The plan the new President Kennedy inherited, one of the greatest blunders in CIA history, was to "invade" with an "army" of fourteen hundred moderately trained Cuban exiles. They never got off the beach where they landed. It was such a massive fiasco that some believed the CIA did it intentionally to force Kennedy into a full-scale military intervention. Khrushchev fired off an aggrieved telegram and rattled his largely fictional ICBMs.

Kennedy had been mortified twice in less than a week. Desperate for a way his administration could get some prestige back, he turned to space. Advised by NASA administrator James Webb, a brilliant bureaucratic strategist, Kennedy decided to announce a race to put a human on the Moon by the end of the 1960s. It would be a genius publicity coup. A race to the Moon was one that the Americans, with their limitless funds and mighty industrial base, were most likely to win.

But first NASA had to prove to Kennedy that they could even put a man into space. On April 25, two weeks after Gagarin's flight, NASA tested another unmanned Mercury on an Atlas booster. The hope was that it would orbit the planet twice before splashing down. As it ascended, the rocket failed to roll and pitch. It just kept going straight up. At forty seconds, the ground crew blew it up. Because it was straight overhead, it rained smoking debris straight down. Terrified television crews and reporters ran and crawled under their vehicles for safety. It gave a whole new definition to "duck and cover."

Finally, on May 5 a Redstone rocket squirted astronaut Alan Shepard out of Earth's atmosphere. It was a suborbital, 15-minute blip of a flight 116 miles up, barely crimping space. It was not nearly so impressive as Gagarin's Earth-striding, but the Americans hyped it as though Gagarin had never been born. If no great triumph, it was at least a good pretext for Kennedy. On May 25, he announced the Moon race to Congress.

It's one of history's ironies that Richard Nixon, whom Kennedy had beaten to the White House partly on the basis of his false "missile gap," actually approved of a Moon race years before Kennedy announced the idea with such great fanfare. In 1958, when the Eisenhower administration was seeking ways to counter Khrushchev's Sputnik coup, the idea of a race to the Moon was floated in a high-level closed-door meeting. Nixon loved it. He saw it as one way to power up his presumed presidential run in 1960. Eisenhower shrugged it off, to Nixon's eternal frustration. Eisenhower thought, maybe correctly, that a Moon race was too expensive and pie-in-the-sky to propose in 1958. But also, he didn't like Nixon any more than, well, anyone did (except Elvis, but he was on many drugs at the time), and was never convinced that Dick Nixon could be "presidential," so he didn't go out of his way to help him campaign. So Kennedy beat Nixon, then made great political hay out of an idea Nixon had supported three years before him.

Khrushchev's response was classic cheek. He sent a diplomat to the White House to present Kennedy's daughter Caroline with a gift from the Soviet people, a fluffy white puppy that was a daughter of Strelka. Kennedy did not miss the message: Game on.

7

SHEE-UT

B ecause American media, propaganda, and pop culture focused so intently on American astronauts, Americans had, and still have, little idea what a global phenomenon Yuri Gagarin became. Khrushchev had his propaganda machines turned up to eleven. Ukrainian shepherds notwithstanding, around the world Gagarin was cheered and thronged like Space Elvis. Except in America, where the Kennedy administration decided that the best way to deal with the Soviets' historic achievement was to act like it never happened. He toured dozens of countries and always drew large crowds. Yurimania predated Beatlemania. Women around the world declared him the sexiest man alive. In Sweden a pretty blonde dashed through a crowd, threw her arms around his neck, and kissed him. The paparazzi spread the photo worldwide, fanning the flames. He flew to London, where he was driven in a Rolls Royce convertible with YG1 license plates to have lunch with

Queen Elizabeth at Buckingham Palace. Looking at all the cutlery
lined up beside his plate, he admitted to her that he didn't know
which he was supposed to use. Charmed, the Queen said, "You
know, I was born in this palace, but I still get mixed up." She also
flirted with him a little, and told him that all the young ladies of
London were swooning over him.

In Ian Fleming's novel *On Her Majesty's Secret Service*, published two years later, a Secret Service analyst named Leathers
tells M and Bond that SPECTRE's Ernst Stavro Blofeld is using
deep hypnosis for "malignant" purposes, then notes that "the
leaders in this field, ever since Pavlov and his salivating dogs,
have been the Russians. If you recall, sir, at the time of the first
human orbiting of the earth by the Russians, I put in a report on
the physiology of the astronaut Yuri Gagarin. I drew attention
to the simple nature of this man, his equable temperament when
faced with his hysterical welcome in London. This equability
never failed him and, if you will remember, we kept him under
discreet observation throughout his visit and on his subsequent
tours abroad. . . . That bland, smiling face, sir, those wide-apart,
innocent eyes, the extreme psychological simplicity of the man,
all added up, as I said in my report, to the perfect subject for
hypnosis. . . ." Leathers concludes that "in his space capsule, Gagarin
was operating throughout in a state of deep hypnosis." Maybe it was
something Fleming heard from his extensive connections in the
intelligence community.

It was rumored that the Italian sex bomb Gina Lollobrigida
sneaked into his hotel room in Rome to spend a night with him.
They both denied it, but they admitted to admiring each other. He
began to manage to have his shirt off frequently when the paparazzi

were around; small though he was, he worked out to stay buff. His goodwill trip to Africa generated rumors about his sexual conquests there. Back home, even the Marxist-Leninist ladies of Moscow got caught up in Yurimania. Hair salons invented a sculpted new hairdo called the Love Me Gagarin. It was reported that girls on the collective farms hoarded newspaper pages with his photo on them and covered them with kisses.

He was the first cosmonaut to have a car of his own. A grateful government gave him a black GAZ Volga, the frumpy 1950s-style sedan you never wanted to see on your block because it meant the KGB were in the hood. This was no fit ride for an international star, and he soon ditched it for a blue Matra Bonnet Djet coupe that was a gift from the French aerospace and automobile manufacturer. The Djet was a hot sports car, something like a French take on the Jaguar XK-E, and nothing like any other Soviet citizen drove. It was his version of the free Corvettes that Mercury astronauts got to race around in. He loved to bomb around in it, scaring the hell out of his passengers, cosmonauts and fighter jocks included, especially when he was drunk and took his hands off the wheel at top speed, laughing like a maniac.

Set loose in a world of wine, women, silver cutlery, and fast cars, Gagarin went from good Soviet peasant lad to bad boy rock star in record time, proof of the power of celebrity to make even a Communist into a Keith Moon. Historian Andrew Jenks points out that it was a phenomenon Soviet society had not quite seen before. The "cult of personality" had insured that Lenin and Stalin were worshipped like secular gods; some individual Soviets, like Kamanin, were famous, at least inside the Soviet Union. But the USSR had never produced a rock star and sex god.

It's no wonder he let it go to his head. Literally. In October 1961, after touring the world all summer, Gagarin went for a vacation at the beautiful Black Sea resort of Foros, a playground for Soviets of the highest rank on the very southern tip of Crimea. He brought Valentina and their little daughter. Titov, Leonov, and some other cosmonauts tagged along, and Kamanin was there to play the KGB chaperone. They were put up at a deluxe sanatorium. It was soon obvious that the boys had no intention of being chaperoned—not by the general and not by their wives. The whole group of cosmonauts partied like fighter pilots letting off steam, scandalizing Kamanin and frightening the sanatorium's nurses with their drunken horse-play. Gagarin ignored Valentina while he caroused with the others, reducing her to tears.

On their last night there Gagarin stalked a pretty blonde nurse to her bedroom, where he sloppily grappled with her. There was a knock on the door, and Valentina called his name. Gagarin went out to the small balcony to escape, and either jumped or fell over the railing. Luckily for him it was only a six-foot drop, but he fell flat on his face, smashing his forehead against a rock. His comrades joked that he hurt himself more falling six feet than he had falling from outer space. Kamanin tut-tutted in his diary that Gagarin "was a hair's breadth from a very nonsensical and silly death." The doctors who stitched him up were sworn to secrecy. He was all right, but his forehead was permanently dented, and when he made his next public appearance he was wearing a fake left eyebrow because they'd shaved the real one. The gash ran down into it, and it never grew back right; in all his photos afterward, the top of his face looks dinged and lopsided. He wasn't the first or last to be scarred by fame, but it isn't always so obvious. There was some discussion of sending him to plastic surgeons in the West, plastic surgery not being much of a thing in Marxist-Leninist society, but the higher-ups nixed that, not wanting the West to get wind of the Hero of the Soviet Union's ridiculous and scandalous escapade.

Two cover stories were generated. In one he tripped carrying his baby girl and, holding her up to save her, banged his face. An even more heroic story appeared in the newspaper *Izvestia*, which had him diving into the Black Sea to save a drowning child and hitting his head on a submerged rock.*

If Gagarin was a bad boy, Gherman Titov was badder. Following Gagarin to the stars and to stardom, he still competed with him and seemed determined to outdo him in every way. First, though, he had to survive the KKK—Korolev, Kamanin, and Khrushchev.

After Gagarin's flight in April 1961, the Soviets didn't put another cosmonaut into space until August. It wasn't that they couldn't. They just hadn't planned for it, and, as surprised as the rest of the world was by their spectacular success, they spent the next few months arguing among themselves about what to try next. It's a classic example of the difference between NASA's careful march forward and the Soviets' haphazard leaps and bounds. In July, Korolev was still arguing with Kamanin and others about how long the next flight should be. Korolev wanted another bold leap, a full twenty-four-hour mission. The others, still worried about

---

* There were two major state-run newspapers, *Pravda* (Truth) and *Izvestia* (News). The running joke was that there was no izvestia in *Pravda* and no pravda in *Izvestia*.

the effects of long-term weightlessness, wanted a shorter flight. Khrushchev, as he often did, stepped in and shoved them forward. He asked Korolev if the next mission could happen by August 10, without explaining why that date was significant. Korolev agreed, and got his twenty-four-hour flight approved in the bargain.

On August 6, when Titov took off, Khrushchev's thinking became clear. He sent word that day to the East German government to begin building the Berlin Wall. He hoped that a triumphant second spaceflight would be a happy distraction from that fraught and dangerous gambit. Actual construction of the wall began the following week, on August 13.

Titov remembered that on the night of Sputnik he'd been thrilled by the thought, "Maybe man can fly in space someday, maybe in 20 to 25 years." Now here he was, less than four years later.

His day in space started out well enough. After the deep pangs of envy he'd felt when Gagarin was picked to fly first, he was giddy with joy. Using his code name, he kept whooping "I am Eagle! I am Eagle!" as he sailed overhead. At one point he said to anyone listening, "I wish you had it so good." When he flew over Canada, he exchanged radio greetings with Gagarin, who was still touring the world. Gagarin joked that he'd wave out a window so Titov could see him.

More worryingly, the prolonged weightlessness that Kamanin had been concerned about did start making Titov feel "seasick." He felt like he was upside down, which made him "giddy and nauseous," and at one point he briefly fainted. Over the years, half of all space travelers have experienced this "space adaptation syndrome." The inner ear needs time to adjust. Despite his queasiness, Titov tried to eat his programmed meals, squeezing pâté, pureed vegetables, cold coffee out of tubes—then throwing them up. He developed a headache, muscle aches, blurred vision. At one point he fell into a deep, exhausted sleep, only to wake up and be startled by the sight of his arms drifting weightless in front of him. For a split second he forgot he was in space. On top of all that, after a failure in the environmental system the temperature dropped to -10 degrees Celsius (-14 degrees Fahrenheit).

Titov dutifully reported all his problems—and his career suffered for it. Some people in the program believed that it showed "a lack of character" for him to say anything. Cosmonauts, like fighter pilots, weren't supposed to admit when they felt ill. They were supposed to man up and shut up. Because he admitted to having so many problems on this flight, his mental and physical fitness for future missions was called into question.

After twenty-four hours, he was well ready to come home. Then his real problems started. His reentry was as rough and potentially deadly as Gagarin's had been. Once again the capsule and the service module behind it failed to disconnect. Korolev had been so busy arguing with others that he hadn't fixed the problem that had nearly killed Gagarin. Once again, luck intervened when the superheated air of reentry burned through the straps and the two modules separated. Disobeying orders, Titov left the porthole windows open and watched "a raging purple flame" turn the glass yellow. It was the heat shield melting around the capsule like candle wax. Korolev's team hadn't solved *that* issue yet either.

Titov safely ejected. As he drifted under his parachute he recognized the broad Volga River coursing through collective farm fields. A locomotive was speeding along a rail line and he had to struggle with his chute a bit to avoid getting run over by it. He landed in a field six hundred feet away from the tracks, and not far from where Gagarin had touched down. Never the best parachutist in the cosmonaut corps, he hit the ground awkwardly, fell backward, and was dragged by his chute. It didn't dent his natural cockiness. On his way to his post-flight briefing with the program doctors, he had a bottle of beer. It struck them as "arrogant."

They had no idea. Already full of himself and a handful for Kamanin, he really cut loose now. Kamanin's diary for the next few years contains an endless string of rumblings and grumblings about Titov's drunken exploits, and the KGB's file on him grew fat with reports of transgressions any one or two of which would have gotten an ordinary air force officer cashiered. Worse, Titov and Gagarin now egged each other on like Keith Moon competing with Keith Richards for debauchery honors. They were so open and proud of themselves for all the sex they were getting that they were brought before a closed-door session of Communist Party leaders, space historian Asif Siddiqi writes, "at which they were warned in no uncertain terms about the grave consequences of negative publicity." It made no impression. They had both survived near-death experiences in Korolev's rattletrap space machines. They were pretty sure that made them invulnerable. Titov later said that he did start cutting back on the sex—not because of the party's threats, but because he just couldn't handle so much. There's no indication that he cut back on the vodka. It's no wonder Kamanin recorded suffering from insomnia.

It must be remembered that before they were world heroes, before they were jet pilots, these two were country lads poor as potatoes, growing up in a ruined land, survivors of the worst the twentieth century had to throw at them. Their idea of a high time had been an extra glass of vodka. Now the world with all its hot cars and hot women was laid out before them. And they knew there was little their higher-ups could do to punish them for indulging in it all. After presenting them to the world as great heroes of the people, the government could hardly reprimand or punish them in public. It could only fall back on its familiar mode of cover-up and lies to keep the world from learning that its two great heroes were also great philanderers and drunks.

Gagarin had barely recovered from his silly and nonsensical fall when he and Titov were partying again. Kamanin received a report that they'd gotten drunk at a hotel in Moscow and Gagarin had gotten from his room to Titov's by wobbling along the ledge outside, five stories up. In his diary he noted how much Gagarin loved to drink and predicted he wouldn't slow down anytime soon. Titov was involved in no less than three drunk driving accidents by February 1962, including one where he slammed his car—a humble Volga, not a French sports car, but still—into a bus, and was very lucky no one was hurt. A few months later he cracked his car up in another accident, left the scene, and hid out in Kyiv. He did his own drunken skirt-chasing as well, trying to get his chauffeur, a young mother, to come to his room for drinks at 2:00 a.m. Another time he left folders of top secret papers on the seat when he parked his Volga. Reprimanded for it by a security officer, he waved him off with, "I'm Titov. Who are you?" Drunk on a tour of a factory, he was similarly high-handed with the security guards there. In a motorcade in Romania, he jumped out of his limo and onto a cop's motorcycle.

Inevitably, other cosmonauts began acting out. They sometimes showed up terribly hungover for morning briefings because Gagarin and Titov had kept them up all night with endless rounds of vodka toasts. One of them beat his wife when she confronted him about all the sleeping around he was doing. Called before his superiors, he said he was only following the lead of Gagarin and Titov. He was drummed out of the cosmonaut corps anyway.

So it had to be with some trepidation that General Kamanin prepared to accompany Titov in the spring of 1962 on his most important goodwill tour yet: a visit to the United States. Kennedy decided that it would just look like sour grapes if America continued to shun heroic cosmonauts. Besides, NASA had managed to orbit John Glenn that February, which helped boost American confidence. He only did three orbits compared to Titov's seventeen six months earlier, but once again you wouldn't know that the way America cheered it. So let the Russki come. Maybe it would help ease some of the tension Khrushchev's Berlin Wall was generating.

Titov arrived in New York on April 29, 1962, accompanied by Tamara, with Kamanin hovering. After a couple days of taking in the bright lights, big city, they flew to Washington. The Titovs spent two days there, led around by John Glenn and his wife, Annie. At the Washington Monument, Titov showed off his English. Asked what he thought of the monument, he embarrassed the straight-arrow Glenns by drawling, "Shee-ut! We got obelisk in South Moscow 1,500 meters." *Shee-ut* was evidently his favorite English word and he used it often during the trip, though there's no record of his saying it when he met President Kennedy and Vice President Johnson at the White House. The Glenns invited the Soviets to their home in suburban northern Virginia for an all-American barbecue. Glenn struggled with the grill, and somehow, just as the visitors' limos drove up, the carport caught on fire. It was a small blaze and they soon put it out. "Tell me," Titov said to Glenn, "every time you have a barbecue you burn the house?" A member of Titov's entourage remembered it as "a long evening filled with burned steaks and superb Russian vodka." Despite that mishap Kamanin was very impressed with Glenn, who struck him as the sort of clean-cut, responsible family man he wished Titov and Gagarin were.

From there they flew to Seattle for the World's Fair. Big crowds surrounded Titov and were reported to be surprised he was so friendly and charming. At the NASA pavilion he saw a replica of Glenn's Friendship 7 capsule and, doing his part for Khrushchev's agenda, blandly lied that it was tiny compared to the "spacious" Vostok. The next day, the Titovs had lunch at Trader Vic's, then headed back to the fair, with a gaggle of reporters dogging their every step and Kamanin at Titov's shoulder. A reporter asked Titov if his spaceflight had affected his philosophy of life. Speaking through his interpreter, Titov said, "Sometimes people are saying that God is out there. I was looking around attentively all day but I didn't find anybody there. I saw neither angels nor God." Jaws dropped. "Up until the orbital flight of Major Gagarin," he continued, "no God was helping make the rocket. The rocket was made by our people and the flight was carried out by man. So I don't believe in God. I believe in man—in his strengths, his possibilities, and his reason."

He didn't say *shee-ut*, but what he did say fell like a bomb. Newspapers all around the country, previously charmed by him, now condemned him as a godless Communist insect. Editorials spilled barrels of ink thundering about Soviet "anti-religious propaganda." It's not clear whether any of them noticed that he was paraphrasing a quote from the Futurist poet Vladimir Mayakovsky's "The Flying Proletarian": "We've inspected the sky inside and out. / No God or angels were detected." It was a sort of slogan for the space program. Anyway, the uproar didn't stop American teenyboppers from mooning at him. As with many movie stars, Titov's dashing good looks, wavy blond hair, and bad boy attitude qualified him as a sexy teen idol despite his small stature and his married status. Back in New York City for the last leg of the tour, he did a fawning interview with *Seventeen*. At a Ford factory in New Jersey, two teenage girls off to one side made goo-goo eyes at him. One of them held up a handmade sign: TITOV—YES! USSR—NO!

It wasn't just American teenagers enamored of Titov, either. Kamanin records that gaggles of girls in the Cosmodrome town of Leninsk would stand in the rain for hours outside the cosmonauts' quarters and chant Titov's name to try to coax him out. Little of this made any sense to the upright old Stalinist.

As celebrities, Titov and Gagarin weren't just ladies' men. They were manly men's men as well. A lot of the partying and horsing around they did was for males only—other cosmonauts, their KGB escorts, others inside the circle. Gagarin loved organizing trips out to the forest to hunt wild boar and bear. Jenks reports that "A monument in the forests around Kaluga marks the spot where he once bagged a moose. . . . The monument reads: 'In these forests on October 20, 1967, the first cosmonaut in the world and hero of the Soviet Union, Yuri Alekseevich Gagarin, participated in recreational hunting.'" Sometimes, like LBJ, he hunted partridge from a moving vehicle. He liked to pile a bunch of guys, fishing tackle, and some cases of vodka onto a boat and go caroming out onto a lake. They called it fishing, but as the supply of vodka dwindled his piloting grew faster and more erratic, and the only way to catch a fish would be if you fell overboard and came up with one in your mouth. He started a water-skiing club, frowned on by older, stiffer sorts as a bourgeois leisure activity.

Some of Gagarin's restless acting out came from boredom. He had been the first human in space, and basked in the adulation for that. He wanted to go again. He wanted, in fact, to be the first human to set foot on the Moon. He thought he deserved it. Despite his recent gaffes, he was still Sergei Korolev's favorite of the little eagles. Korolev made it no secret that he'd be happy to oblige his young friend. Sick, broken, frequently a nervous wreck, Korolev wasn't sure how much time he had left to achieve his ultimate *mechta* of beating the Americans to the lunar surface.

Putting Yuri Gagarin on top of another one of Korolev's frankly hit-or-miss rockets was the last thing the Soviet political leadership wanted. He was their greatest propaganda asset since Sputnik. Why risk him? They didn't even want him to pick up his career as a jet pilot, let alone a space jockey. Let some other young hero go. Yuri Gagarin would best serve the Soviet people with his feet planted on terra firma.

So in effect they grounded him. Without officially saying he'd never put on a spacesuit again (although they did that later), they gave him a desk job, appointing him deputy director of cosmonaut training at Star City, reporting directly to Kamanin. To soften the blow, they also promoted him to colonel. Star City became his private fiefdom, his Graceland, where he and his friends lived and partied the way only the tiniest fraction of Soviet citizens could. With Gagarin wielding his star power, the amenities at Star City got fantastically luxurious by Soviet standards. The commissary chef "trained in culinary arts at the Hotel Metropol, the Soviet Union's most prestigious restaurant," Jenks writes. "The cooking staff kept its own garden for fresh produce as well as cows, chickens, and pigs. 'Plates with vegetables and salad were always in the dining hall,' remembered the cook to the cosmonauts. Cosmonauts could come into the hall at any time for a snack. 'The most startling thing,' remembered one new Star City denizen in the 1960s, 'was fresh tomato on my plate—in the middle of April!'" While ordinary citizens would stand in line for hours for the slim chance of buying a new coat or pair of shoes, loads of new clothes were laid out in the gym for cosmonauts to grab for free. The only cost to them was walking out with their arms full past the sad eyes of the trainers and janitors. Gagarin got Star City denizens access to the best medical and dental care available at the time.

At his most ambitious, he oversaw the design and construction of new housing for the cosmonauts, trainees, and their families. No more cramming two families into one tiny apartment or sleeping in a corner of the gym. Gagarin insisted on state-of-the-art living quarters, with mind-boggling three- and four-bedroom apartments. Four bedrooms! For a single family! Muscovites crammed into Khrushchev's grim brutalist apartment blocs, hastily thrown up and instantly dilapidated, didn't believe the rumors they heard of these palatial digs in a hidden city just forty kilometers away. Gagarin also had a new clubhouse built, called Cosmonauts' House. "It has been the hub ever since of social, cultural, and public life in the city," Jenks writes, "the venue for parties, festivals, receptions, concerts, and countless banquets and feasts in which many a bottle of vodka has been downed and rendition of 'Moscow Nights' performed." Gala annual parties there, like New Year's Eve and Cosmonautics Day (April 12, the anniversary of Gagarin's historic flight, still celebrated), drew the crème de la crème of Soviet elites and oligarchs.

The very best caviar, vodka, and music available anywhere in the Soviet Union were on hand. Gagarin usually played the toastmaster, and the toasts, each supposed to be accompanied by a shot of vodka, could go on until the wee hours of the morning, when there was no one left sober enough to lift a glass.

Of course the cosmonauts knew this was all the height of hypocrisy in a Communist society where all citizens were supposedly equal. But the elites and oligarchs had lived this way long before these young men joined them at the trough. Most cosmonauts had grown up in extreme, stone-hard poverty. Was it difficult for them to avoid the janitors' hungry stares? Probably, but not impossible.

Gherman Titov was not content to do his drinking only in Star City. He still caromed around the countryside, drunk and dangerous. In 1964 he got a dressing-down from the commander of the air force for his insubordination and lack of discipline, and seems to have shrugged it off. One night that June he was out for another drunken joyride and picked up a female hitchhiker. Speeding wildly down a dark country road, he had another one of his crashes. His passenger was injured and unconscious, but he was miraculously unharmed. He was too drunk to ever give a full account of what happened, but somehow he caught a taxi home, leaving her there. She died in an ambulance on the way to the hospital. When police identified her they discovered she had an eight-year-old son.

Kamanin had had more than enough. He wanted Titov drummed out of the cosmonaut corps and the party, never again allowed to drive a car much less pilot a spacecraft or aircraft, and stripped of all rank, privileges, and titles. But he also knew none of this would happen. He wrote glumly in his diary:

> Titov is not only known by the Soviet people; the whole world knows him. The disgrace of Titov will be the disgrace of all cosmonauts, the disgrace of our people. We cannot allow this and will make one final attempt to keep Titov in the party and in the cosmonaut team.

A closed-door military court cleared Titov of all charges. The government privately arranged to have the dead woman's son admitted to a prestigious military academy.

8

THE COSMONETTE

Bang! Zoom! You're going to the Moon!
—RALPH TO ALICE KRAMDEN, *THE HONEYMOONERS*

In 2019, the *New York Times* ran a remarkably naïve article titled "How the Soviets Won the Space Race for Equality." The writer claimed that by putting not only the first woman into space but also—many years later—the first Asian and first black man, the USSR signaled in unambiguous ways its "commitment to equality."

Yeah, no. As *The Atlantic* retorted, "The Soviet Space Program Was Not Woke." The truth is that the first woman in space, Valentina Tereshkova, got to go because, once again, the Soviets were worried that the Americans were going to beat them to it. It was another one of Khrushchev's publicity stunts. Had the Soviets known the truth—that NASA was way, *way* far from even entertaining the idea of female astronauts, much less letting one of them go to outer space—it's highly unlikely that Tereshkova would have flown either.

To be fair, the Americans sent some confusing signals. The cover of the February 2, 1960, *Look* posed champion airplane pilot Betty Skelton in a silver spacesuit in front of a Mercury capsule and asked, SHOULD A GIRL BE FIRST IN SPACE? If the Soviets actually read the article, they should have realized that the answer was a resounding NO! At *Look*'s request, NASA had put Skelton through some of the same testing as Mercury 7 astronauts, who nicknamed her Mercury 7 1/2. It was for the magazine article, not a serious move toward training women astronauts. In her book *Almost Astronauts*, Tanya Lee Stone notes that American women at the time couldn't rent a car or take out a loan without a male cosigner. They weren't allowed to fly fighter jets, let alone spaceships. One NASA official snarled at Skelton, "Women in space? If I had my way I'd send them all out there!" Cue Ralph Kramden.

Still, a NASA doctor named William Lovelace began thinking, why not? Women were generally smaller and lighter than men, both pluses, and there was no lack of highly skilled women pilots.

Women had been flying since the early years of aviation, and Rosie the Riveter pilots had performed essential duties during World War II. None of them were combat fighter jocks, but only because there was a law against it. On his own dime Lovelace recruited twenty-five women pilots to go through the astronaut training he conducted for NASA at his Lovelace Medical Center in New Mexico. He subjected them to the same grueling physical and mental tests male astronauts, and cosmonauts, suffered through. Thirteen of the twenty-five made it. Years later they'd be hailed as the Mercury 13, but NASA never seriously thought of taking them on, and this was as far as their "astronaut training" went.

Nevertheless, when General Kamanin visited the US with Titov in the spring of 1962, he met one of the Mercury 13, ace pilot Geraldyn "Jerrie" Cobb, and somehow got the impression that NASA was well on its way to putting a woman in space. In his journal he wrote, "Under no circumstances should an American become the first woman in space—this would be an insult to the patriotic feelings of Soviet women." When he returned to Russia he discussed it with Korolev, who took some persuading. They spoke to Khrushchev. Khrushchev, characteristically, decided that if the Americans were going to put women in space, the Soviet Union should do it first.

Word went out to flying and skydiving clubs around the USSR, which were very popular and had many female members. Applications flooded in from some eight hundred women. Under Kamanin's direction fifty-eight were selected and went through rigorous physical and psychological testing. In the end it came down to five. They all had extensive parachute training. One of them had made nine hundred jumps. One was an avid pilot who was getting a degree in mathematics. One was a cosmonaut-trainee's wife.

The one who emerged as the star, Valentina Tereshkova, wasn't a pilot, but skydiving was a passionate pastime of hers. Like so many male cosmonauts, she was born stone poor, in an unheated wooden hut on a collective farm on the Volga. Her father was a tractor driver who was drafted into the Red Army when she was two and promptly killed driving a tank in the Winter War in Finland in 1941. Her mother worked in a textile mill, and Valentina joined her. She was always sporty and adventurous, swimming, skiing, bareback riding, but her mother was still unhappy when she signed up at a local skydiving club with a female friend from the mill. On her maiden jump she stepped out of the plane before her startled instructor had given her the signal. When they were both safely on the ground he screamed at her. It was not the last time she betrayed a kind of hardheadedness that was not quite insubordination but close enough to aggravate her superiors.

Meanwhile, she became secretary of the mill's Komsomol committee for good Communist youth, a public relations plus for a cosmonaut. She later remembered the elation of listening to the news about Gagarin's flight and then Titov's on the mill's radio. After Titov's she impulsively sent a letter to the Supreme Soviet in Moscow offering herself as a cosmonaut candidate.

Kamanin brought his five female candidates to Star City for training. The question was, training for what? Korolev, with just two Vostok missions behind him, Gagarin's and Titov's, wanted to jump boldly ahead yet again. Even though their near-death experiences showed that there were a lot of issues with the Vostok capsule still to be solved, Korolev wanted to put two or even three Vostoks into orbit at the same time. Not just a juggling act, coordinating their flights would be a step toward docking and undocking spacecraft, which was a step toward putting a cosmonaut on the Moon.

Then Khrushchev surprisingly ordered him to pause the program. The military was clamoring for the Zenit spy satellite Korolev had been promising them since 1959, and Khrushchev now directed him to forget all else and concentrate on that. The Cold War was heating up. There was the CIA's dismal Bay of Pigs misadventure in April 1961, and Khrushchev's decision to begin building the Berlin Wall that August. That same month the Kremlin said it would resume atmospheric nuclear testing. The US and USSR had tested nukes furiously up until 1958, when they agreed to an informal moratorium. Now both sides started detonating them in the atmosphere again, including the Soviets' fifty-megaton Tsar Bomba in October 1961, still the largest nuclear explosion in history, three thousand times the bang of the Hiroshima bomb.* The Americans responded with a pretty big bang of their own. They were concerned that the electromagnetic pulse (EMP) from Soviet nukes detonated in space could confuse or even destroy their ICBMs. To test this notion, they put a 1,400 kiloton warhead on a Thor missile and triggered it in space, 240 miles above the Pacific near Hawaii. The results were spectacular. An aurora like a giant bruise in the sky was visible for thousands of miles around. The EMP was far stronger than expected. It blew out traffic lights and phone lines all over Hawaii, and fried at least six satellites, including the famous Telstar 1, making it a victim of the space-nuke version of friendly fire.

Khrushchev and his generals wanted spy satellites because they knew the Americans were mapping Soviet territory practically inch by inch with their own very successful one, code-named CORONA. That's not an acronym for Collecting Orbital Reconnaissance ... or such. It was just a name a project planner came up with as he gazed down at his Smith-Corona typewriter.

---

* They dropped it over the Novaya Zemlya archipelago in the Arctic, the Bikini Atoll of the frozen north, which they nuked an astounding 131 times over the years. But the most-nuked region on the planet is a zone of Kazakhstan called by the appropriately sci-fi name the Semipalatinsk Polygon. From the very first Soviet atom bomb test in 1949 to 1989, no fewer than 483 nuclear devices were detonated there. Some 1.5 million people were exposed to the fallout, then monitored like lab rats to clock the effects. Related illnesses and abnormalities including genetic mutations persist to this day.

Presumably calling the project SMITH would have been a serious breach of flair. The all-caps was a CIA convention. The first successful CORONA (after a dozen failed tests) was launched right after Eisenhower halted U-2 flights over Soviet territory in 1960. In fact it completed its first mission on August 18, 1960—the very day that Francis Gary Powers was sentenced in Khrushchev's kangaroo court in Moscow. That day CORONA returned more images of Soviet territory than the twenty-four previous U-2 missions combined. In a single mission it proved that, far from cranking out ICBMs like sausages, the Soviets had maybe half a dozen of Korolev's tetchy R-7s. Six months earlier, American intelligence had guesstimated that the Soviets might deploy two hundred nuclear-tipped ICBMs in 1961. Now they dropped that to twenty-five, tops.

That CORONA worked at all was something of a miracle. It was definitely an example of the Wrong Stuff American-style. It has been called "the world's most expensive disposable camera." A two-stage Thor-Agena rocket would launch in top secrecy from Vandenberg Air Force Base on the California coast. It would vault the satellite into low orbit. As Soviet land rolled by below, CORONA's panoramic camera (later two cameras, for stereoscopic imaging) snapped pictures. The exposed film was fed into a "bucket" that was sealed in a small reentry capsule. At a signal from ground control the satellite jettisoned this capsule, which then hopefully glided into the assigned reentry trajectory. At an appropriate altitude its parachutes (hopefully again) opened and it drifted down toward the Pacific near Hawaii. An air force C-130 from Hickam Field, with a specially trained crew, trailing long drag lines, would (hopefully a third time) swoop in and snag the reentry capsule *in midair*. This actually worked more often than it sounds like it should have.

Why this complex procedure? Why not just send up television cameras and let them beam back their images? The Soviets had in fact been testing video in space since 1959; when Luna 3 took its photos of the far side of the Moon it used a television system to transmit them. A television camera inside the very first Vostok beamed down video of Gagarin. Why not CORONA? Because the resolution of television was far too low. Footage from both the Moon and inside Gagarin's Vostok was grainy and blurry. It would be useless for intelligence purposes. Film was the only option.

Korolev and his team came to the same conclusion for Zenit. He was pulled away from developing his manned (much less womanned) program for a full year while working on the satellite. Its first twelve launches failed, including two of his skittish R-7s blowing up so spectacularly that serious damage to the launchpad caused frustrating delays. In July 1962, the same month the Americans were fritzing satellites with their EMP wave, a Vostok/Zenit was finally successfully launched from the Cosmodrome, took pictures of the continental United States, and parachuted back down to Soviet soil with its film. For the first time, the generals had some hard targets in the US at which to aim their ICBMs (such as they were). Before this, had they launched, they would have been firing blindly—not that they told Khrushchev that. Before July 1962, Sergei Khrushchev wrote, "When he brandished missiles, Father had no idea that the military simply didn't know where to shoot." Now the generals were well pleased with Korolev for once. Their only complaint, Chertok notes, was the low quality of the Soviet film used. They wished they had some high-grade Eastman Kodak film like CORONA did. Still, over time, Zenits and other surveillance spaceware would provide the Soviet military with startlingly accurate and detailed maps of the US and elsewhere—more accurate than American maps in some instances—which only came to light in the 2010s.

While Korolev and his team were thus employed, the five would-be female cosmonauts spent a year at Star City, effectively a fighter jocks' secret clubhouse, ruled over by the star, Yuri Gagarin. Offended and intimidated by the idea of letting women in, Gagarin and the boys were rude and condescending to them. Gagarin called them "little tarts." He saw no use for them as cosmonauts. Neither did Titov, though he did see another way they could be useful: as Kamanin gloomily notes, he once convinced two of the five to spend a night with him in his bedroom. The women went through all the grueling physical and mental training the guys did—the centrifuge, the vibration seat, the dreaded isolation chamber. They were given some rudimentary pilot training, even though, since Korolev was still using the Vostok capsule, knowing how to pilot a plane would be a far less useful skill for them than knowing how to eject from one. Their trainer was a much-decorated combat pilot, Colonel Vladimir Seryogin. Tereshkova said he was "a kind and patient teacher." We'll meet him again. Tereshkova gradually emerged at the top of the pack—Kamanin called her "a Gagarin in a skirt." Khrushchev was said to like her as a good working-class heroine, Soviet style.

In August 1962, having delivered Zenit, Korolev was able to resume the Vostok program. Taking another leap ahead of the Americans, he put two capsules in orbit simultaneously. Before the launches, the Soviets submitted an unusual request to the US, which said in effect: *Please don't explode any more nukes in space right*

now. You'll endanger our cosmonauts. The Americans complied. Vostok
3 carried Andrian Nikolayev, who despite looking like a bulldog
sniffing a fire hydrant was apparently romancing Tereshkova. (There
weren't many lookers in the cosmonaut corps, of either gender.
Gagarin's and Titov's unusually celebrity-ready mugs, at least before
Gagarin broke his, were definitely a factor in their being chosen
as cosmonauts No. 1 and 2.) Vostok 3 was launched on August 11.
Korolev's ground crew then scrambled madly and got Vostok 4 off
the ground from the very same launchpad less than twenty-four
hours later. It carried Pavel Popovich. At one point during the mission Nikolayev and Popovich drew to within three kilometers of
each other. Some in the press hailed this as a "rendezvous in space."
Astronaut Wally Schirra, who carried out the first true rendezvous
in space three years later, correctly sniffed that this was more of "a
passing glance."

In terms of publicity it was an EMP blast all its own. The world
press called Nikolayev and Popovich the "Heavenly Twins." A
Soviet propaganda poster showed four sleek, sci-fi-looking rockets
soaring away from Earth and declared "Gagarin, Titov, Nikolaev,
Popovich—the mighty knights of our days." (If there was one area
in which the Soviets outdid NASA every time, it was in beautiful
propaganda posters.) NASA administrators, their astronauts, and
President Kennedy himself were staggered by the Soviets' audacity. It
didn't help that just a week earlier the *Saturday Evening Post* had run
an opinion piece by former president Eisenhower blasting Kennedy's
Moon program as "a mad effort to win a stunt race." As the Twins
continued to circle over their heads—and wave to them in grainy live
TV feeds—NASA programmers glumly speculated that the Soviets
might be preparing to attempt docking the two vehicles. NASA was

years from trying this, and had assumed the Soviets were too. They were right. In truth, the Vostoks were only as crudely maneuverable as ever. They had simply been launched so that their orbital paths brought them near each other. But the Americans didn't know that, and, as Vasily Mishin recalled years later, the Soviets sure weren't going to enlighten them: "As they say, a sleight of hand isn't any kind of fraud. It was more like our competitors deceived themselves on their own." With the Americans still reeling, the Soviets were poised to rock them once more by putting the world's first female in space. And then the Cold War intervened *again*.

Now that Khrushchev knew CORONA had shown the Americans the true puniness of his ICBM arsenal, he was more worried than ever that they might try a first strike. Unable to count on those ICBMs, he devised a plan to use smaller IRBMs—intermediate-range ballistic missiles—to make the Americans think twice before they attacked. IRBMs in Soviet territory didn't have the range to be a threat to the US. Khrushchev got Fidel Castro to agree to let him deploy some in Cuba, where they could hit Washington, New York, and other sites across the continental United States. He referred to it as putting a hedgehog down the Americans' trousers. The IRBMs were products of Yangel's shop. Korolev was reaching for glory and had no interest in medium-range rockets. Before he gave them Zenit, the military men used to say, "Yangel works for us. Korolev works for TASS," the Soviet news agency. The missiles' deployment in Cuba was done in secrecy. How Khrushchev planned to frighten the Americans with secret missiles was a bit mysterious.

In a classically inane Soviet ruse, the program was called Anadyr, the name of a small Arctic settlement, and the ships carrying the nukes were piled up with skis and sheepskins. Somehow the Americans saw through this charade when a U-2 photographed the Cuban launch sites under construction in October 1962. With his typical false bravado, Khrushchev claimed he was surprised it took so long for the Americans to spot them—it "showed the weakness of American intelligence," Sergei later said. Kennedy reviewed his options, rejected Curtis LeMay's request to bomb the hell out of Cuba, and went with a naval blockade. Once again, Khrushchev had maneuvered himself into a tight spot, one that this time might trigger a nuclear holocaust.

Meanwhile, with manned flights from the Cosmodrome put on hold yet again, Chertok and Voskresensky were innocently supervising the launch preparations of a Mars probe. Having put the first probe on the Moon, the Soviets were eager to beat the Yankees to other planets. Korolev was home in Moscow, supposedly down with a cold, though it might well have been nervous exhaustion. Suddenly the army flooded the Cosmodrome with submachine-gun-wielding soldiers "in full combat gear, even wearing gas masks," Chertok recalled, "in case of attack by US paratroopers." He and Voskresensky were ordered to get their modified R-7 off the pad so that the military could replace it with another rocket carrying a nuclear warhead. The country might be about to go to World War III. Chertok explained that defueling and moving a tetchy liquid-propellant R-7 would take many hours if they didn't want it to explode in their faces. And that the Mars launch window was just a few days long. If they aborted the mission now they'd have to wait two years to try again. But the military had their priorities. Chertok managed to get through to Korolev, who said he'd speak to Khrushchev. Then he and Voskresensky sat down in the Gagarin cottage to wait, cutting up a large watermelon and toasting each other from two bottles of ice-cold vodka. "May God grant that this not be our last drink!" was one of Voskresensky's. Chertok remembered hearing a joke while this was happening. A listener calls in to Soviet Armenia's state radio, Radio Yerevan, and asks, "What should one do in the event of a nuclear missile attack?" Armenian Radio replies, "Quickly roll yourself up in a white sheet and slowly walk to the cemetery." "Why slowly?" the listener asks. "So that you don't cause a panic."*

Finally, the military men were ordered to stand down. Chertok and Voskresensky were allowed to launch their rocket. It turned into yet another classic Wrong Stuff event: the rocket blew up as it was ascending, causing radar operators at the US Ballistic Missile Early Warning System to think a Soviet nuclear attack had begun. Luckily for all, they hesitated before pushing any big red buttons and actually commencing World War III.

On October 27, a Soviet SAM missile battery shot down a U-2 over Cuba, killing the pilot. Curtis LeMay redoubled his efforts to get Kennedy to approve nuking Cuba out of the water, but both Kennedy and Khrushchev recognized it was time to step away from the

* Radio Yerevan was so ridiculously heavy-handed in the way it obediently pumped out Kremlin propaganda that it became the butt of many jokes in the Soviet Union, and has continued to be in post-Soviet Russia. In one of the most famous from the 1980s, a listener asks if there is freedom of speech in the USSR like there is in the US. Radio Yerevan answers yes: In the US you can stand in front of the White House and shout "Down with Reagan!" and not get arrested, and in the USSR you can stand in Red Square and shout "Down with Reagan!" and not get arrested. In 2022, Putin's invasion of Ukraine prompted a whole new round of Radio Yerevan jokes.

brink. They communicated privately, and Khrushchev began with-drawing the Cuban missiles.

Finally preparations could get under way for Tereshkova's flight. It was decided that it would be another "twin" launch. Vostok 5 would carry cosmonaut Valery Bykovsky. Tereshkova would fly in Vostok 6. Yuri Gagarin was the celebrity guest at a Cosmodrome ceremony before the launch. He quipped that it was nice to see so many friends and coworkers gathered together "approximately at Baikonur." Making a joke about Tyuratam's cover story was the sort of comment only a star would risk.

Technical glitches, one after another, caused the launches to be delayed and delayed again through the first two weeks of June. Tempers flared as nerves were frayed raw. "Korolev had been seriously ill in recent weeks," Siddiqi writes. "He had a fever for several days and was diagnosed with inflamed lungs. He looked 'pale and wan' to everyone, his voice hoarse from talking." Bykovsky finally climbed into Vostok 5 on June 14. Then he sat in the capsule six hours while Korolev yelled as one small problem after another popped up. During that time, like Alan Shepard before him, Bykovsky was over-whelmed by a need to empty his bladder. Finally, there was a power cable that wouldn't detach. Korolev looked like he was going to have a coronary. Voskresensky characteristically said screw it, it will detach itself on launch. And that's what happened. As the R-7 rose from the pad, everyone sighed with relief and went outside to light another horrible Soviet cigarette. No doubt Korolev and Voskresen-sky gulped a couple of their Validol pills.

Tereshkova lifted off from the Cosmodrome in Vostok 6 two days later. Unfortunately for the prospects of the female cosmonaut program, her flight was Wrong Stuff all the way. She and Bykovsky only got to within five kilometers of each other, farther apart than Vostoks 3 and 4, and could not even spot each other as little dots against the stars. Not even a passing glance. She got as space sick as Titov had, struggled with the manual attitude controls, skipped other programmed chores. She complained about the lousy cosmonaut food, and threw some of it up. As her wobbly flight dragged on, her communications with the men in ground control, including Gagarin and Korolev, grew abrupt and testy. She complained about how uncomfortable her helmet and pressure suit were, considering she had to wear them for nearly three days. She didn't like that for cleaning herself she could only run a moist towelette over her face, and the towelettes weren't moist enough. Mishin would claim that she "turned out to be at the edge of psychological stability." It was later rumored, inevitably, that at least some of her discomfort and crankiness was because she was having her period in low-g.

Despite all that, Tereshkova spent more time in space, just shy of three full days, and made more orbits, forty-eight, than all the American astronauts put together at that point. She remains the only woman to fly a solo space mission. Still, the problems continued even as she was ending it. During reentry she alarmed mission control by not reporting normally on her progress at first. When she did, her voice sounded strangely listless as she deadpanned, "The ship is turning, turning quite fast, turning, starting to burn. In [my] field of vision I see the burning ship. Such reddish light, reddish. The ship is turning and burning. Like a pendulum it's turning and burning, burning. It's swinging, swinging, burning in

[my] field of vision. It's burning vigorously. It's burning vigorously. Swinging around the axes, swinging around the axes. It's shaking, it's shaking. Crackling." That she reported all this in such a dull, resigned-sounding monotone spooked listeners on the ground. Even after she ejected and her parachute opened, she had one more problem. As she drifted to the ground she looked up at her chute, and a piece of metal that had fallen off the capsule banged her on the bridge of her nose. She covered up the bruise with extra makeup when she met Khrushchev.

Bykovsky had his own reentry problems. Korolev's team had still failed to address the problem of the capsule and the service module not separating when they should. Like Gagarin and Titov before him, Bykovsky tumbled wildly until the cables burned through. He and Tereshkova both ejected safely and came down not far from each other, but far north of the target zone, in the remote and mountainous Altai Krai, where Titov grew up. They had to wait overnight for the retrieval teams to reach them. Tereshkova didn't mind. She was very happy to be back on the ground and out of her pressure suit, and whiled away the night with the locals. Journalist/propagandist Evgeny Riabchikov reported that when the retrieval team and official media arrived, "The cameramen filmed Tereshkova drinking fresh milk from a bottle as she stood beside her spacecraft in the steppe. People from neighboring kolkhozes [collective farms] surrounded the metallic sphere and peered in through the hatch. They asked Valentina about her flight and offered her cheese, lepeshki [flat dough cakes], kumiss [fermented mare's milk], and bread. From every direction, herdsmen on horseback came galloping up to see the spaceship."

An ecstatic crowd filled Red Square to cheer Tereshkova. They laughed when Khrushchev mischievously shoved Tereshkova and Nikolayev together and they sheepishly embraced. Hearing somehow that they had dated, he apparently decided the first cosmonaut wedding would be yet another publicity victory. Insiders said the two were friendly but not *that* friendly, and Kamanin, at Khrushchev's urging, had to spend months hectoring them every day to set a date. They finally caved and tied the knot in November at the "Griboyedov Palace of Marriage." It was just a registry office. A grinning Khrushchev gave the bride away. In the "cosmic wedding" photos Tereshkova and Nikolayev still don't look *that* friendly. They look like a middle-aged pair stiffly going through the motions, standing near each other but not quite together. Champagne and caviar flowed at the reception, where Khrushchev and Gagarin toasted the couple no fewer than twenty-one times. Khrushchev, ebullient and braggadocious as ever, had invited the Moscow correspondents of foreign press outlets to the wedding. One of them was Theodore Shabad of the *New York Times*. Shabad noticed two men lurking at the back of the crowd and did some digging. His article in the November 12 issue of the *Times* named them: Korolev and Glushko. It was the first time they were named in Western media. The Soviets just shrugged and mugged when asked about them.

The newlyweds tottered off to a new home gifted them by the proud Khrushchev: a *seven-room* apartment. Tereshkova gave birth to a daughter by Cesarean seven months later, indicating that the couple was friendly at least once before the wedding. They would become estranged and finally divorce in 1977.

Around the world Tereshkova's flight was hailed as a great leap forward for women. Except, not surprisingly, in America. The response in America was largely to wave it off as a publicity stunt. One newspaper dismissively referred to her as a "cosmonette." Astronaut Scott Carpenter said that a chimp could do what she did. That a chimp could have done what he did too apparently didn't enter his mind. *Life* magazine quoted an unnamed NASA guy, possibly the Ralph Kramden one, saying that the very idea of female astronauts made him "sick at my stomach." A rare note of dissent came from journalist-politician Clare Boothe Luce. Writing in that same issue of *Life*—her husband Henry's *Life*—she used Tereshkova's triumph as a way to shame the men at NASA and American males in general for their Kramdenesque attitudes. Although she and Henry held anti-communist views, she was also a feminist. She argued that Tereshkova's flight was an example of the "spectacular" progress women had made under Soviet communism. Precisely the lesson Khrushchev wanted the West to take, which the *New York Times* was dutifully echoing as late as 2019.

As usual, no one outside the program knew of Tereshkova's difficulties during the flight. To the press she was upbeat about the experience. But Korolev and others on the inside were disappointed with her performance. Kamanin defended her, and the female cosmonaut program generally, but Korolev drew the conclusion, as Leonov later wrote, "that his spacecraft were unsuitable for women." Leonov questioned "the Soviet policy of strictest secrecy" that hid the truth of Tereshkova's performance from the public. "Most Russians believed that women should not meddle in what was considered men's work. When Tereshkova returned to Earth and talked about how simple the mission had been, our [the male cosmonauts'] hearts sank. The prestige of the cosmonaut corps sank a little, too."

Now that he'd won his propaganda points, Khrushchev lost all interest in the cosmonettes. Tereshkova's version of the glittering world tours Gagarin and Titov got was to be sent to places like Ghana and Lebanon, and the logistics were terribly botched, most likely sabotaged by the men doing the planning. She returned to Star City and directed the training of the other cosmonettes, but no missions were scheduled and the program was quietly terminated in 1969. Tereshkova remained in the air force and would retire as a major general in 1997. She also had a long career in politics. Still staunchly patriotic, she was a member of the Presidium from 1974 until the Soviet Union was on the brink of collapse, then held various parliamentary positions in the post-Soviet Russian Federation.

The next Russian woman in space, Svetlana Savitskaya, didn't get there until 1982. Still, that was quicker than the Americans, who didn't get around to putting a female in space until the following year, with physicist Sally Ride's mission on the space shuttle *Challenger*. It took sixty years for one of the Mercury 13 to finally go into space—sort of. Mary Wallace "Wally" Funk was eighty-two when she rode as a guest on Jeff Bezos's Blue Origin in July 2021. The ten-minute flight peaked at about seventy-six kilometers, shy of the one hundred kilometers conventionally considered the edge of space, so technically none of the Mercury 13 had gone into space still.

Tereshkova was also eighty-two in 2021, and also in the news again. In 2007, Vladimir Putin had hosted celebrations of her seventieth birthday. As a politician, she was a stout supporter of Putin and his nostalgia for the glory days of the Soviet empire, of which, after all, she was a living avatar. She could be seen in the audience when Putin gave his interminably droned speeches, her grandmotherly hair helmet rigidly coiled and swooped. In 2021 she proposed an amendment to the Russian constitution that would allow Putin to evade term limits and remain in office until 2036. It passed. Putin's opponents in Russia denounced her as a puppet who helped him crown himself dictator for life. "I don't even want to talk about these people who don't love the country," she shrugged.

9

SPACE GHOSTS

"There is an eerie possibility that a long-dead Russian astronaut is today hurtling silently through space at thousands of miles an hour—the victim of a Soviet space shot that went wrong," *Reader's Digest* reported in 1965. "His body perfectly preserved by intense cold, he may be a lonely wanderer in space for centuries to come."

Wait, *Reader's Digest*? That's a bit grim for a family magazine mostly read at the dentist's office, isn't it? Even at the height of the Cold War. But then, dead cosmonauts were all over Western media at the time, triggered by Yuri Gagarin's shocking achievement. It's worth pausing a moment here to consider this phenomenon, which said as much about the Soviets as it did about the Americans.

While most of the world was cheering the Soviets for Sputnik and Gagarin and Titov, others in the West, mostly but not only in the US, had questions the Soviets were not answering. They didn't know the true extent of Soviet lies and cover-ups, but they knew enough to doubt that the Soviets were conquering space quite as effortlessly as they pretended. Stories of cosmonauts whose deaths in space had been covered up started to circulate as early as 1958, after Sputnik 1 and 2, and then ran rampant after Gagarin's and Titov's flights. They showed up in fringe sources like UFO newsletters and sci-fi magazines, but the *Digest* was far from the only mainstream venue to indulge. Collectively, they came to be known as the Lost Cosmonaut stories.

The science fiction writer Robert A. Heinlein started one of the most repeated. He was staunchly anti-Communist and anti-Soviet. Although the gung-ho kill-'em-all militarism of *Starship Troopers* was satirized in the 1997 film adaptation, Heinlein seriously believed that it was how the West should reply to Soviet aggression. He just used hordes of sentient arachnids as stand-ins for Communists.

In October of 1960 the conservative journal the *American Mercury* ran a long Heinlein article with the title "'Pravda' Means 'Truth.'" He reports that he and his wife, who was fluent in Russian, happened to be in the Soviet Union the previous May when Gary Powers was shot down.

> About noon on Sunday, May 15 we were walking downhill through the park surrounding the castle which dominates Vilno. We encountered a group of six or eight Red Army cadets. Foreigners are a great curiosity in Vilno. Almost no tourists go there.
>
> So they stopped and we chatted, myself through our guide and my wife directly, in Russian. Shortly one of the

cadets asked us what we thought of their new manned rocket?

We answered that we had had no news lately—what was it and when did it happen? He told us, with the other cadets listening and agreeing, that the rocket had gone up that very day, and at that very moment a Russian astronaut was in orbit around the earth—and what did we think of that?

I congratulated them on this wondrous achievement but, privately, felt a dull sickness. The Soviet Union had beaten us to the punch again. But later that day our guide looked us up and carefully corrected the story: the cadet had been mistaken, the rocket was not manned. . . .

This is the rocket the Soviets tried to recover and later admitted that they had had some trouble with the retrojets; they had fired while the rocket was in the wrong attitude.

So what is the answer? Did that rocket contain only a dummy, as the pravda now claims? Or is there a dead Russian revolving in space?

I am sure of this: at noon on May 15 a group of Red Army cadets were unanimously positive that the rocket was manned.

There *was* a cosmonaut on board, and he was a dummy: Ivan Ivanovich. On May 15, 1960, Korolev managed for the first time to put a Vostok capsule in orbit. It was known in the West as Sputnik 4. It was a crucial step toward orbiting a man a year later, with Ivan sitting in for the future human cosmonaut. On May 19 mission control sent Sputnik 4 a command to fire its retro-rockets for reentry.

Instead of sending it toward the ground they pushed it into a higher orbit. So Heinlein and the cadets weren't entirely wrong, only maybe half wrong: it was a manikin, not a man, up there circling, circling.

Over time Sputnik 4's orbit decayed and it came down of its own accord, in two parts, the vehicle and its instrument module. The vehicle, and evidently Ivan, broke up and burned on reentry over Wisconsin on September 5, 1962. People saw what looked like a streaking meteor and heard a sonic boom. A twenty-pound chunk of it landed in the middle of a street in the Lake Michigan town of Manitowoc. It must have been the biggest thing to happen in Manitowoc (population about thirty-five thousand) in years, because to this day the town celebrates Sputnikfest every September. The instrument module disintegrated on reentry in 1965.

As soon as Gagarin flew, the Moscow reporter for the London *Daily Worker* ran with the rumor that a Colonel Vladimir Ilyushin, test pilot son of the aircraft designer Sergei Ilyushin (whose Il-14 had so embarrassed Khrushchev in 1955), had already secretly flown three orbits in a spacecraft called the *Rossiya* (Russia) a week before Gagarin. It's possible the reporter got this from a Moscow cab driver—they were notorious sources of both fact and rumor well used by foreign press. According to the story, Colonel Ilyushin's time in space had left him "in a bad way," both physically and mentally. French newspapers repeated the tale. The perfectly respectable American magazine *U.S. News & World Report* went so far as to claim that Gagarin never flew, he was just used as the handsome, photogenic stand-in for the damaged Ilyushin so that the Soviets could claim a triumph instead of failure.

Colonel Ilyushin *was* in fact "in a bad way"—but from a car accident, not space madness. He healed and continued a long career as a test pilot, founded the Soviet Rugby Federation in his spare time, and died an eighty-two-year-old retired major general in 2010. Yet the rumor persisted as late as 1999, when it was the subject of a one-hour documentary—or "crockumentary," as one space historian put it—called *The Cosmonaut Cover-Up*.

Colonel Pyotr Dolgov was another who was rumored to have flown in space, and died there, months before Gagarin's flight. A writer of flying saucer books claimed that stations around the world tracked the fatal flight in October 1960. They did not. Dolgov *did* actually die for the space program, but not in space, and not until after Gagarin flew. In 1962 the Soviets were sending experienced parachutists up into the stratosphere using high-altitude balloons. They were testing the Vostok ejection system and pressure suits. Dolgov held a world record in parachuting. On November 1, 1962, he made his 1,409th jump, from a Vostok-shaped balloon gondola that was 93,970 feet up, wearing a pressure suit. His parachute opened, then the pressure suit totally failed. He died quickly, but his corpse took thirty-seven minutes to drift to the ground. As usual the government covered up the true cause of his death and it only went public after the fall of the Soviet Union.

The government also lied about a failed Venus probe that only got as far as Earth orbit, where it stalled and every tracking station on the planet registered it. The Soviets pretended it was a "heavy satellite" that was *meant* to be stuck in Earth orbit. Observers noted that it was big enough to carry humans, and yet another rumor about circling cosmonaut corpses was launched. (The probe was equipped with a flotation ring that would be activated when it reached Venus. At the time, astronomers still believed the surface of Venus might be covered in oceans.)

In October 1961, two months after Titov's flight, a big solar flare supposedly knocked a three-man Soviet capsule out of orbit and sent it off into the interplanetary wasteland. When Titov visited the United States in 1962, he met the famed newspaper columnist Drew Pearson in Washington. Pearson grilled him about this and other Lost Cosmonaut rumors. Titov annoyedly waved him off.

Then there was the sad case of Grigory Nelyubov. He was one of the top early cosmonauts, an excellent fighter pilot, considered a shoo-in for one of the first missions, always mentioned along with Gagarin and Titov. Titov didn't like Nelyubov breathing down his neck and apparently used his influence as the corps' second-brightest star to keep Nelyubov down. He was the number two backup for Gagarin on Vostok 1, first backup for Titov on Vostok 2, and backup again for Vostoks 3 and 4, but still didn't get a mission of his own.

The lack of advancement made him bitter. One night in 1963 he and three other cosmonauts got drunk and rowdy in the restaurant at the little train station near the Chkalovsky air base, two miles off the Star City campus and off-limits to cosmonauts, whose presence in the area was supposed to be secret. "The more they drank the more argumentative and abusive they became," Leonov recalled. "Eventually they picked a fight with one of the restaurant's waiters and the stationmaster called the local military base." When military police showed up to collect them, Nelyubov got abusive with them, too. Titov was filling in at Star City for Gagarin, who was doing a publicity tour. Titov offered Nelyubov a chance to apologize. Nelyubov replied that the MPs could "go to hell" and refused. All four of the offending cosmonauts were cashiered out of the corps and sent back to the air force.

Nelyubov and one of the others, Ivan Anikeyev, continued to spiral to oblivion. According to Leonov, Anikeyev was drunk at a party when someone stole his car keys, hit and killed a pedestrian, and slipped the keys back in Anikeyev's pocket. Anikeyev was imprisoned for a year before he was declared innocent and released; his career as a pilot was over. Nelyubov was stationed at an air base near Vladivostok, which is to say, nowhere. Miserable, he appealed to Kamanin and Korolev to let him back into the corps, with no luck. No one at his air base believed that he'd once been a cosmonaut. Depressed and drinking heavily, he experienced what one Russian journalist called "a crisis of soul." One snowy night in 1966 he was walking along some railroad tracks when a train hit and killed him. The state ruled it suicide. His widow agreed.

Because Anikeyev and Nelyubov were considered embarrassments to the cosmonaut program, the government pulled a trick it had been performing since the Stalin era: they became, as Orwell put it in *1984*, unpersons. Their names were struck from written records, and they were airbrushed out of cosmonaut group photos. One minute they're standing or sitting there, grinning at the camera alongside Gagarin and Titov and Korolev. The next minute, *poof!* They don't exist. They never did. The Sovietologist James Oberg noticed the disappearing acts in photos from the 1960s. In one famous shot from 1961, Korolev posed with Gagarin, Titov, and five other cosmonauts, with Nelyubov standing behind Korolev. In that same photo from a few years later, Nelyubov has vanished. Another well-known shot from the same period shows Korolev surrounded by the whole cosmonaut training corps. In a version the Kremlin circulated a few years later, no fewer than *five* of those cosmonauts have disappeared. In 1986, enjoying the loosening of the reins on journalists in the Soviet Union's last few years, Golovanov went public with the Nelyubov story for the first time, writing in *Izvestia*. By then all sorts of rumors had circulated about Nelyubov and the other mysterious disappearing cosmonauts. It was also Golovanov who first clued outsiders in to Valentin Bondarenko's horrific fiery death.

One of the most disturbing Lost Cosmonaut tales was started soon after Gagarin's flight by a pair of ham radio operators near Torino in the northwest corner of Italy. In the 1950s the teenage brothers Achille and Giovanni Judica-Cordiglia took over a concrete bunker the Germans had left behind at the end of the war. They filled it with scavenged equipment, festooned it with two dozen antennae, and the Torre Bert Listening Station was born. In 1957/8 they tracked Sputnik 2 and Explorer 1 passing overhead. In 1960, they said, they tracked an object slowly moving away from Earth, its transmissions of SOS in Morse code gradually fading into silence. They were convinced it was a secret Lost Cosmonaut whose vehicle had gone off track and was drifting away. In February 1961, they thought they could hear the labored breathing of an orbiting spaceman suffocating to death. Then in May 1961, just a month after Gagarin's flight, they claimed that they recorded an unidentified female cosmonaut whose mission was going very badly. In the chilling tape, which they made public, a frightened-sounding Russian woman cries, "I see fire! I see fire! I feel hot. Am I going to crash? Conditions getting worse. Why don't you answer? Please!" Was she in a Vostok capsule burning up on reentry? Was it a hoax? The Soviets recorded no manned (or womanned) flight for that period. The first female cosmonaut that we know of wouldn't fly for another two years. The Cordiglia brothers' recording, despite being taken seriously by *Reader's Digest* and other venues, was written off as a hoax capitalizing on all the Gagarin hoopla. But it's still hair-raising. (It's online.) The Soviets denounced the brothers and reputedly sent a KGB man their way once with a warning. Kamanin dismissed them with the memorable slur "gangsters of space." In 2007 an Italian documentary about the brothers appeared, called *I pirati dello spazio—Space Pirates*, but translated into English as *Space Hackers*.

The typical Soviet reaction to Lost Cosmonaut tales was to play the aggrieved victims of "enemy" slander. Even Golovanov, who wrote the truth about Nelyubov and Bondarenko, toed the party line, denouncing the tales as "a well thought out anti-Soviet propaganda campaign . . . to belittle our country's scientific and technical achievements." No it was not. Like other unhappy outcomes, the Soviets brought this on themselves. Their habit of keeping secrets and telling lies about their space program begged for skepticism. It was impossible to believe that they were soaring from one victory to another without the usual human quotient of mistakes and errors, so outsiders filled in those blanks. After the dissolution of the Soviet Union in 1991, a flood of documents and crowds of eyewitnesses took up the mantle from the conspiracy theorists. Now everyone knows much more about the Soviet space program, good and bad, than anyone except a handful of insiders and Soviet officials knew at the time. But we still don't know how much we don't know, and we never will. There are surely hosts of secrets from back then, certainly about military operations in space, that will remain buried forever.

And so, as rumors will, Lost Cosmonaut tales persisted long after the Soviet Union itself vanished. On April 12, 2001, Reuters celebrated the fortieth anniversary of Gagarin's flight with a story about three men who supposedly died in suborbital flights launched from Kapustin Yar from 1957 to '59. The source of the story had been an engineer on Valentin Glushko's team. His story was debunked by both American and Russian space experts, but only after it had already made the rounds.

10

CLOWN CARS
IN SPACE

After Tereshkova's flight in March 1963, the Soviets put no one else into space for almost a year and a half. Korolev and his team wanted to work on a whole new spacecraft, an evolutionary leap forward from the primitive Vostok sphere. Instead, at Khrushchev's insistence, they were sidetracked for six months figuring out how to cram three hungry men into a one-man Vostok and call it a Voskhod. As we know, they miraculously managed to put that contraption into orbit in October 1964.

When the orbiting cosmonauts spoke to Khrushchev at his dacha, Anastas Mikoyan, who had become chairman of the Presidium just a few months earlier, was standing next to him. At the end of his conversation with the trio in space, Khrushchev said that Mikoyan wanted to speak to them too. "He is literally tearing the receiver out of my hand, and I cannot deny him." No, of course he was not. Khrushchev was no Stalin, but you still wouldn't dare to tear the phone out of his hand. Khrushchev was horsing around, as he so often did when he was feeling bubbly.

When the Voskhod trio landed and pried themselves out of the smoking capsule on October 13, they expected to be in Moscow on October 15 for the traditional Red Square celebration and meeting with Khrushchev. But a funny thing happened on their way to the Kremlin. On October 14, Khrushchev was removed from office. "He is literally tearing the receiver out of my hand, and I cannot deny him" were the last words he ever spoke in public.

As usual, it was his own fault. He had dealt his and the Soviet Union's prestige a severe blow with his Cuban missile debacle two years earlier. The state propaganda apparatus played his self-defeat as a great victory, of course; confronted by the brash young American hothead, he had shown restraint and wisdom and saved the world from oblivion, etc., etc. Privately, though, the members of the Presidium muttered darkly. They'd been putting up with his mercurial and sometimes foolish behavior for years, and this time he almost got them and millions of other people killed. A cabal of them, with Leonid Brezhnev as their somewhat reluctant figurehead, began to plot how to shuffle him out of the way. Other Communist leaders around the world agreed it had been a fiasco. In China, Mao shook his head at the foolish "adventurism" that had ended with "capitulation-ism." In Cuba, Khrushchev biographer William Taubman relates, Castro raged, calling Khrushchev an "asshole" and a *maricón* with *no cojones*. In private, Kennedy concurred. "I cut his balls off," he snickered.

No one was more aware of the scope of his failure than Khrushchev himself. "Khrushchev had learned at last that bluff and bluster didn't pay," Taubman writes, "but they had been his main weapons, and without them, he was lost."

Still, paradoxical as ever, Khrushchev came away from the Cuban missile crisis liking missiles even *more* than he had before. Not medium-range missiles like the ones that had just gotten him into trouble, but big, nuke-delivering ICBMs. In fact, he was more convinced than ever that they were the key to the Soviet Union's safe, secure future. In March 1963 he brought all the generals and admirals together and told them that traditional armies and navies were obsolete. From now on war would be nuclear, fought with ICBMs. He was going to cut deeply into their budgets for ground and naval forces and reallocate, he said. If he wanted to push them into the conspirators' camp he couldn't have done a better job.

As though he knew how short his time was getting, he grew increasingly restless and cranky, manifested in trivial but telling ways. When a censored version of Aleksandr Solzhenitsyn's Gulag novel *One Day in the Life of Ivan Denisovich* was published in Moscow in 1962, Khrushchev astonished everyone, including Solzhenitsyn, with his enthusiastic praise for it. But hopes for a cultural thaw were dashed soon after when he went apoplectic at an exhibition of avant-garde art. "It's dog shit! A donkey could smear better than this with his tail." He said one painting looked like "a child has done his business on the canvas . . . and then spread it around with his hands." He railed at an artist, "If that's supposed to be a woman, then you're a faggot. And the sentence for them is ten years in prison."

When he went to Pitsunda for a holiday the conspirators made their move. They'd sent Mikoyan with him to keep him occupied until they had worked out all the details. Mikoyan was a good friend and trusted ally to Khrushchev—they had next-door mansions in Moscow—but he agreed it was time for Khrushchev to go. On October 14 Khrushchev responded to an urgent summons to return to Moscow. There he was driven from the airport straight to the offices of the Central Committee of the Communist Party. All 169 members of the committee sat there staring at him as he walked in. It was said that his old supporter Brezhnev could barely look him in the eye as one accuser after another stood and denounced him. All behind closed doors. No show trial, no firing squad.

Muscovites and foreign press took note the next morning when an immense banner of Khrushchev's face, three stories tall, which had gazed down over the Kremlin from high up on the Moskva Hotel, vanished. They noticed there were no photos of Khrushchev, no quotes from him, no mention of him at all in *Pravda* or *Izvestia* on October 15. *Pravda* finally ran a short piece the next day, claiming that Khrushchev had resigned due to "advancing age and deteriorating health." Without naming him, an accompanying editorial condemned "harebrained scheming, half-baked conclusions and hasty decisions and actions divorced from reality." If you didn't know it was about him, you would when you got to the bit about "bragging and bluster." Brezhnev took over. Now it was his face in the papers, heavily retouched to conceal that he looked like a thousand-year-old lizardman.

As Khrushchev had done with his defeated opponents, Brezhnev et al. allowed him to live, as long as he did it quietly. In effect, he was pensioned out. His ride was downgraded from a boss's Zil to a middle manager's Volga. He lost both the lavish Pitsunda dacha and his palatial mansion in Moscow's Lenin Hills. His new dacha in the Moscow suburbs was more modest than either of those, but very nice, with a garden, woods, and waterfront on the River Istra. His neighbors were government ministers. His favorite pastime was to wander out on the property and start a bonfire, which he'd stare into for a long time. Sometimes he tried to fish in the river, but if anyone ever lacked the patience for fishing, it was Nikita Sergeievich.

After removing him from power, the government proceeded to remove him from history as well. For instance, after 1964 it was as if he'd never met the country's greatest celebrity. "New editions of Gagarin's autobiography edited out all mentions of Khrushchev," Andrew Jenks writes. Among all the many photos in Yaroslav Golovanov's *Our Gagarin*, published in Moscow in 1978, there's not a single one of Khrushchev. A photo of Gagarin on the Lenin mausoleum is severely cropped to show only him, not Khrushchev beside him. On the next page it's Brezhnev pinning a medal on Gagarin. If you didn't already know what a huge role Khrushchev played in Gagarin's flight, you wouldn't know it after leafing through this massive tome, either. It was only after the Soviet Union collapsed that the real story and photos reappeared.

In 1966 he began dictating his memoirs into a tape recorder, sitting out in his garden, no doubt because he was certain the house was bugged. It's believed he took this risky move because of how completely he'd been erased from official history. He wanted his side on record. It took a few years. Sergei hid copies of the transcript in various spots, and got one to the US. In 1970, when Little, Brown announced the upcoming publication of *Khrushchev Remembers*, the Kremlin called the old man in for a shouting match. He accused his accusers of being tsarists and Stalinists, called one of them a "little idiot," demanded the death penalty, then signed a "confession" that the book was a fabrication. He died of a massive heart attack the following year. He was seventy-seven. *Pravda* printed a very short, bland death notice the next day, the first time his name had appeared in its pages in seven years. Given that he'd been one of the two most powerful men in the world, his funeral was stridently down-market. His successors made certain of it. No Red Square ceremony, just a quick wake in a morgue on the outskirts of the city that none of his former colleagues attended. The weather was appropriately dreary and overcast as his casket and his family were driven in an old bus to Novodevichy Cemetery, more or less Moscow's Père Lachaise, known for the graves of Gogol, Chekhov, Prokofiev, and other cultural figures. Not as honorable a final resting spot as the Kremlin Wall, but not so bad for the former farm boy. Soldiers kept rubber-neckers out. A brass band played the national anthem as he was lowered into the ground, the small gathering tossed in flowers, and everyone dispersed to get out of the rain and away from the lurking KGB notetakers.

The Voskhod trio waited for several days at the Cosmodrome while Brezhnev settled into power. While there they gave the Soviet version of a press conference: public figures lying to fake news media. Komarov dutifully told the whoppers that Voskhod was "far more comfortable" than Vostok and that their landing was so feathery soft the vehicle made no impression on the ground. Finally they were summoned. Their reception in Red Square was the first time

Brezhnev appeared in public as the new leader of the USSR. After-ward, Komarov returned to Star City and continued his cosmonaut activities. He would fly again, on another outrageously reckless mis-sion, in 1967. Feoktistov and Yegorov never flew into space again. In 1969, Feoktistov toured the US on a goodwill mission. Accompanied by American astronauts, he visited Disneyland and rode Tomorrow-land's Flight to the Moon ride. He joked that though he never made it to the actual Moon, at least he saw the Magic Kingdom. Yegorov conducted medical research until he died in his Moscow apartment of a heart attack in 1994, at the age of only fifty-seven.

Despite how pointless, dangerous, and slapped-together Vos-khod had been, it had done its job as a publicity stunt. So naturally it was followed by a Voskhod 2. Voskhod 2 continued the Wrong Stuff tradition of being a rushed, impromptu, and nearly deadly gambit to beat NASA to yet another punch. NASA finally announced that its first two-man Gemini mission, Gemini 3 (1 and 2 were unmanned tests), would launch on March 23, 1965. Gemini 4 would follow in June. The plan for that flight was to have an astronaut open the hatch and stick his head and shoulders out, the first time a human emerged from a capsule to the vacuum of space. It would not be a full EVA—extravehicular activity, or what the press decided to call a "space walk." That would come later. NASA was being cautious as always.

Even though Khrushchev wasn't around anymore to encourage zany gambits, the Soviets couldn't resist another chance to upstage the Americans on this one. It was like NASA intentionally teed them up with this half-hearted, head-and-shoulders EVA. The Soviets decided to launch Voskhod 2 a few days before the Gemini 3 flight, *and* to include a real, full-on EVA.

Like its predecessor, Voskhod 2 was a quickly modified Vostok sphere, with its bulky ejection system removed, two seats crammed in it this time, and an improvised air lock added. That air lock was the vessel's riskiest make-do. NASA's plan for a Gemini EVA was uncomplicated. The astronauts would depressurize the capsule and simply open the door. One of them would climb out, do his EVA, then return, close and seal the door, and they'd repressurize the cabin. The Soviets could not depressurize the Voskhod capsule, because *all its on-board electronic systems were old-fashioned vacuum-tube tech that needed to be air-cooled.* If you released all the air from the cabin, the avionics would overheat and fail, and the cosmonauts would die drifting in space. NASA's transistorized equipment presented no such problem. American computers had begun the evolutionary leap from tubes to solid state in 1959. The Soviets were woefully unprepared for the switch to transistors, and fell behind again when Americans started building computers with integrated circuits, which grew ever smarter and more compact. While the greedy capitalists in the West competed with one another to crank out new and improved computers, centrally planned computing in the USSR stagnated at 1940s levels of development. It's a stunning marker of Soviet dysfunction that *no cosmonauts flew with onboard digital computers until 1980.*

So if a cosmonaut was going EVA, Korolev's team was forced to invent an external air lock. They came up with a rubbery tube that telescoped out from the Voskhod sphere. A cosmonaut would crawl out a hatch into this contraption, close the hatch behind him, then open a second hatch at the end of the tube and crawl out to space. He'd crawl back through the tube to reenter the capsule. Like a lot of Wrong Stuff technology, it was as dangerous and desperate as it was ingenious, and it very nearly killed the first cosmonaut who used it.

The same was true of the new EVA spacesuit. Its design was rushed and also all but fatally flawed. And the same could be said again for the extremely accelerated training the two lead cosmonauts, Alexei Leonov and Pavel Belyayev, and their backup team were raced through. The intense stress actually gave one of the backup cosmonauts a heart attack. Meanwhile, unmanned tests went so poorly that the KGB suspected sabotage and placed agents at Korolev's workshop, making his staff really jumpy. One vehicle orbited successfully, but then its parachutes failed on reentry and it hurtled into the ground at such a velocity that even Ivan Ivanovich wouldn't have survived. Another one in February, just a few weeks before Leonov and Belyayev were to fly, blew up in orbit.

Normally you wouldn't launch people in a craft that hadn't passed even a single unmanned test flight, but Korolev was under enormous pressure, some of it self-imposed as always, to beat the Americans again. "Korolev came to see Pasha [Belyayev] and me in the hotel just hours after the [February] explosion," Leonov later wrote. "He looked exhausted and strained. He had not been well: he had been suffering from a high fever as a result of a lung inflammation. *But nothing would deter him from consulting with us and ensuring our safety.*" (Emphasis added, eyebrow raised.) Late that evening he sat down with us and presented us with a stark choice.

"All the data from the unmanned mission has been lost," he told us. "We have only one Voskhod vehicle remaining which is ready for immediate launch. That is your vehicle. If we use this vehicle for another unmanned launch to test the equipment for your space walk, Lyosha, your mission will be delayed by a year until a replacement spacecraft can be built. What is your opinion?" ... Then, very cannily, he added that he believed the Americans were preparing their astronaut Ed White to make a space walk in May. He knew how to get our competitive juices flowing.

Indeed.

On the morning of March 18, 1965, five days before the scheduled Gemini 3 flight, the two cosmonauts opened a bottle of champagne with Korolev and Gagarin; they all took sips, and Gagarin said he'd save the rest until they returned. On the way to the launchpad they got off the bus and pissed on the right rear tire. Then they rode the elevator to the top of an R-7, clambered in through a hatch, and blasted off in an unproven vehicle on a very iffy mission. Even though there were only two men aboard this time, because they were wearing the new spacesuits it was still quite cramped. The liftoff proceeded without problems—which was lucky, since without an ejection system they would have had no means of escape—and they were soon in orbit and ready to try an EVA. While Belyayev inflated the tube, Leonov strapped cumbersome oxygen tanks on his back, enough air for ninety minutes, though the EVA was planned to take much less than that. When the tube was pressurized he climbed up into it, a very tight fit in his bulky spacesuit. Belyayev closed the hatch and depressurized the air lock, then Leonov opened the outer hatch and gave himself a slight push to float free on his tether. An external TV camera beamed the footage home. Suvorov was concerned by how clumsy and jerky the spaceman's movements seemed. Leonov's family was downright distraught. In his memoirs he writes:

When my four-year-old daughter, Vika, saw me take my first steps in space, I later learned, she hid her face in her hands and cried. "What is he doing? What is he doing?" she wailed. "Tell Daddy to get back inside. Please tell him to get back inside." My elderly father, too, was distressed. Not understanding that the purpose of my mission was to show that man could survive in open space, he remonstrated with journalists who had gathered at my parents' home. "Why is he acting like a juvenile delinquent?" he shouted in frustration. "Everyone else can complete their mission, properly, inside the spacecraft. What is he doing clambering about outside? Somebody must tell him to get back inside immediately. He must be punished for this."

The two of them had more sense than anyone in the space program. Floating around in space was much harder work than it looked like. The pressurized EVA suit was stiff as a full-body cast. Just from trying to move in it a little, Leonov's heart was racing and he was pouring sweat. When it was time to climb back into the air lock, he found the suit had ballooned like the Michelin Man. It had expanded so much "that my feet had pulled away from my boots and my fingers no longer reached the gloves attached to my sleeves." He couldn't fit into the tube feetfirst as planned. He flipped around and tried to haul himself in headfirst, but he could only wriggle in part of the way, his legs dangling outside. Exhausted, his suit actually filling up with sweat, his visor fogged, he realized he could die there. So did the people down in ground control, who abruptly cut the television feed. Then, to a blank screen, Mozart's doleful "Requiem" began to play. Soviet viewers—including Leonov's daughter and father—knew that meant trouble. State television did that whenever someone important had died. Years later, Leonov revealed that his helmet was equipped with a suicide pill that he could have taken as an alternative to suffocating when his oxygen ran out. It leaves an indelible image, Voskhod 2 orbiting the Earth with a dead cosmonaut dangling outside.

Korolev and Kamanin had picked Leonov for the world's first EVA for his athleticism and because he was cool under pressure. Once he and his wife were in a taxi that skidded off a wintry road and plunged through the ice in a pond. Leonov saved his wife, then dove back and saved the driver as well. He showed those qualities now. He realized the only way he would get back into the air lock was if he reduced the pressure in the suit. If he depressurized it too far he could get the bends, but there was no choice. It worked. He wriggled in headfirst, then laboriously turned himself around so he could seal the outer hatch. Belyayev let him in.

Mission accomplished, but their troubles had only just begun. As they prepared for reentry, they fired the explosive bolts to eject the air lock. This caused the Voskhod to go into a wild spin they couldn't stop. They asked ground control's advice and got ominous silence. Meanwhile, instruments showed the oxygen pressure in the cabin rising at an alarming rate. They knew that in an over-oxygenated situation a single spark could incinerate them, as had happened to the Apollo 1 astronauts and young Bondarenko. It also suggested that the ship was trying to compensate for an oxygen leak somewhere. "We were both deeply troubled," Leonov writes laconically. Then the automatic reentry guidance system failed. They had to reenter the atmosphere by hand in a malfunctioning metal ball with only rudimentary pilot controls. It was still spinning uncontrollably and still leaking oxygen. The indicators said they had only enough fuel for one attempt. If they failed, their metal ball would skip off the atmosphere and they'd die circling the planet for up to a year before their orbit decayed on its own and they returned as a fireball.

Bizarrely, they suddenly heard a cheery, familiar voice from the ground.

"How are you, Blondie? Have you landed yet?"

It was their friend Yuri Gagarin, calling from a tracking station in Crimea. He evidently had not been tracking very closely. Maybe he'd continued drinking? Blondie was his nickname for Leonov. They told him they were in a life-threatening situation. No time to chat.

To manually point the ship in the right attitude for reentry, Belyayev had to use an optical device called a Vzor and a hand control. Now the problem was that in refitting a Vostok as a two-seater Voskhod the engineers had placed this equipment to Belyayev's left, where he couldn't reach it strapped into his chair or even see it with his helmet on. He unbuckled and lay across the benches, with Leonov holding on to him so he wouldn't float off in the low-g. When they'd strapped themselves back down they fired the retro-rockets—and immediately felt a strong jerk that slammed them into their seats with the force of ten g's. Blood vessels in their eyes burst from the sudden pressure. They were spinning out of control again. "Looking out of my window, I realized with horror what was happening," Leonov remembered. The cables between the sphere and the equipment module were still connecting the two. "It was causing us to spin round its common center of gravity as we rapidly entered the denser Earth atmosphere." *The same problem that had been plaguing*

Soviet missions since it nearly killed Gagarin and Titov. In four years of chasing Korolev's dreams and answering to Khrushchev's agendas no one had paused to fix this problem that they all knew was potentially deadly.

Once again, luckily, the intense heat as they tumbled toward the ground burned through the cables and separated the two units, saving their lives. They breathed again when their main parachute opened and they were drifting quietly toward the ground—and yet another risky situation. Because they had fiddled the reentry by hand, they missed their planned landing site by a whopping 1,240 miles. They came crashing down through snow-covered trees in an isolated forest in the frozen Siberian wilderness. The ball wedged itself between two trees, hanging several feet above the snow-covered ground. Belyayev wondered how long the rescue teams might take to reach them. Three months, Leonov joked grimly, with good dog sleds. Both of them had grown up in Siberia, and they went into survival mode. When they tried to blow the escape hatch, the cabin filled with stinging gunpowder smoke but the hatch didn't budge. "Looking out of the hatch window we could see it was jammed against a big birch tree. We had no alternative but to start rocking the hatch violently back and forth, trying to shift it clear of the tree."

Eventually they clambered out and "sank up to our chins in snow. Looking up, we could see we were in the middle of a thick forest, a taiga of fir and birch. Our main chute fluttered in a tangle above us caught in the branches of trees that must have been 30 or 40 meters tall. The heat of the landing module was rapidly melting the snow and ice beneath it and slowly, before our eyes, it sank onto firm ground." Then the sun set, it began to snow, and they retreated back into the capsule. As their emergency transmitter broadcast their location, they heard the howling of hungry wolves. They knew large bears were out there as well, and it was the mating season. Given the choice between packing extra rations or a pistol, Leonov had packed a standard military officer's 9mm Makarov. He was glad to have it now, though he knew it probably wouldn't stop a hungry bear.

It was late the next afternoon when a civilian helicopter happened to pass overhead and spot them. Soon the sky was filled with well-intentioned helicopters and small airplanes. One dropped a bottle of cognac that broke on impact. Another dropped an ax, and another "two pairs of wolf-skin boots, thick pairs of trousers and jackets. The clothes got caught in branches, but we managed to retrieve the warm boots and pulled them on." Then night fell again, and the temperature plunged to twenty-two below Celsius.

The next morning they discovered that a pack of hungry wolves had gathered around them. A low-flying cargo plane scared them off. A rescue team on skis finally arrived. It took another day to get Leonov and Belyayev back to Baikonur, where Korolev and Gagarin greeted them with hugs.

As usual the government suppressed any public mention of the dangerous malfunctions. As far as the world knew, Leonov's space walk had been another unsullied Soviet triumph in space. Leonov and Belyayev did the usual heroes' press tour of Eastern and Western Europe and Cuba, dutifully keeping their traps shut about any problems. They even managed to lie straight-faced through a lengthy on-camera interview with NBC correspondent Frank Bourgholtzer; the network bumped Bob Hope's regular show to air it that May as a one-hour news special, "The Man Who Walked in Space." The show gave no hint that the man who walked in space nearly died there, because Leonov and Belyayev had done such a great job of snowing Bourgholtzer.

Their experience in the frigid forest convinced Leonov that cosmonauts needed better weapons. He lobbied the bosses for years. In 1982, cosmonauts began packing a specially designed, three-barreled sawed-off shotgun, the TP-82. Because of worries that a cosmonaut might go space-crazy and fire the gun in flight, wreaking catastrophic harm inside the capsule, it was argued that it should be housed in a compartment on the outside of the capsule, where it could only be accessed after landing.

After the collapse of the Soviet Union, the true nature of the accident-prone flight became known, and legendary. Leonov consulted on a film known in English as *Spacewalk* or *Spacewalker*, a Russian equivalent of *Apollo 13*, that came out in 2017.

Embarrassed by Leonov's triumph—and unaware of how close to death he'd come—NASA decided to be less timid for once and go for a full EVA on the next Gemini flight. Gemini 4 took off on June 3. Ed White fully exited the capsule. There was no finicky air lock as on Voskhod 2; White and his copilot James McDivitt simply depressurized the capsule and opened the door. NASA also one-upped the Soviets with a handheld gas gun White used to maneuver around more easily than Leonov had. He had no trouble climbing back into the capsule, though he and McDivitt did struggle a while to close and seal the hatch before the cabin could be repressurized. But that was nothing like Leonov's nearly fatal dilemma. Korolev and his team may have done it first, but the Americans did it right.

11

NOT NOT KNOCKING ON HEAVEN'S DOOR

Too many "chiefs" and not one boss. You can't get
anywhere without a big man in charge.

—SOVIET SPACE TECHNICIAN

On August 2, 1971, astronaut David Scott of Apollo 15 parked
his Moon buggy near a brooding bulb of mountain called
Mons Hadley. He Moon-hopped a short way across the powdery surface. From his spacesuit he drew out a small plaque and a figurine.
The shiny aluminum figurine, three and a half inches tall, was anonymous and sexless, potbellied like a little Botero. Scott lay it in the

sticky lunar dust, and stood the small plaque, not much bigger than a playing card, near it. The plaque listed the names of fourteen astronauts and cosmonauts known to have died by then. In *Two Sides of the Moon*, the 2004 book he cowrote with Alexei Leonov, Scott recalls that he stood there in respectful silence for a brief moment. Then he snapped a picture, hopped on the buggy, and got back to work.

We now know that the list of names could have been much, *much* longer. He'd have needed a plaque the size of a small billboard. It could, for starters, include Bondarenko and Nelyubov, as well as the 165 or more people who died horribly in Nedelin's R-16 explosion at the Cosmodrome in 1960. But no one in the West knew much about those things in 1971.

The sculptor of the little figurine, a Belgian artist named Paul van Hoeydonck, said he wanted it to evoke mankind's heroism, with the statue standing on the Moon and reaching out to the cosmos. But Scott lay it on its back in the Moon dust. It's an image of defeat, not triumph. It looks like it's crying out to the stars, "Help, I've fallen on the Moon and I can't get up!"

That's fitting. The Moon race was well over by 1971. The Soviets had given up even thinking of sending cosmonauts there. The following year, even the Americans would stop going.

Just two years earlier, an estimated 650 million people worldwide had watched with awe and wonder as Neil Armstrong stepped down onto the surface of the Moon. Even the Eastern European Communist countries allowed their people to watch. But not Russia itself. The Kremlin blacked out the live TV coverage there. Only a handful of the elite saw it. Among them were Alexei Leonov and some cosmonauts huddled with other military officers in a Red Army electronic surveillance facility in Moscow, surrounded by TV monitors showing it. Leonov had trained for a Moon mission even as it seemed less and less likely. He now felt "what we in Russia call 'white envy'— envy mixed with admiration." The cosmonauts drank a toast to the astronauts. Of course.

Nikita Khrushchev, seventy-five, forcibly retired dictator, and thirty-four-year-old Sergei were on a leisurely car trip with some friends in Ukraine that night. They stopped near the pleasant-by-Soviet-standards city of Chernobyl. Ten years before the infamous nuclear plant went online, Chernobyl was best known for the camping and hiking nearby. They pulled off the country lane they'd been driving, took a small telescope out of the trunk, and did some rueful Moon-gazing of their own.

Nikita Khrushchev had been one of the few people in the world who might have gotten Leonov to the Moon before Armstrong. Even while he kept distracting Korolev with demands for publicity stunts, Khrushchev encouraged him to continue working on putting a cosmonaut on the Moon. Yet he did it in a classically malfunctioning Soviet way. He approved Korolev's complex and lavish plan for accomplishing a lunar landing by 1967 or '68. Then he turned to one of Korolev's rivals, Vladimir Chelomei, and approved his separate and competing plan as well. Gossips said it was because Sergei had gone to work in Chelomei's design bureau. Finally, absurdly, Khrushchev gave Yangel the nod to work on a third lunar program of his own. *Three* lunar programs? Why? Because that was how things were done in the bankrupt Soviet system. In a society rotten with corruption and almost guaranteed to underperform, you covered your ass by throwing everything you could at a problem or project and hoping *something* worked. It was absurdly wasteful and expensive, in an economy that was already struggling. The manned space program had always gotten by on Korolev's bellowing and micromanaging in lieu of real organizing principles. Now it got more confused than ever, with duplicated efforts and limited funds spread out wastefully. Inevitably, all three programs were rushed and slapdash, and none of them succeeded.

Korolev deserves some of the blame as well. The rock-headed determination and towering ego that had brute-forced him to success after unlikely success now failed him. For his elaborate lunar program he needed a brand-new vehicle and a giant new rocket to shove it to the Moon. The rocket, called the N1 (N for *Nositel*, Carrier), was his answer to the Americans' Saturn V. It was a behemoth, thirty-five stories tall and heavy as a thirty-five-story pile of pig iron, with five stages. Even for engineers used to doing everything the Wrong Stuff way, it was highly problematic. It caused the final blowout between Korolev and Glushko. Glushko wanted to give it engines that used UDMH (unsymmetrical dimethylhydrazine). Korolev was adamant about sticking with his usual liquid oxygen and kerosene fuel. He never forgot the Nedelin UDMH explosion; and besides, by this point if Glushko said black, he'd insist on white. The grumbly animus that had been bubbling up between them for twenty years boiled over. When they took their cases to Khrushchev, Sergei overheard Korolev hiss at Glushko, *"You snake in the grass."* They got to where they weren't speaking at all, then couldn't even be in the same room, and finally Korolev banished Glushko from OKB-1. Glushko went to work with Chelomei on his Moon rocket, the UR-700 (Universal Rocket). It would use UDMH and come to be called the Proton.

Korolev turned to a jet engine designer who'd never worked on rockets—but who also had never dared contradict Korolev. All he could come up with to get the colossal N1 off the ground was a first stage that used no fewer than *thirty* small, not terribly powerful engines. There were extreme difficulties in coordinating multiple engines with the precision needed to prevent a liftoff from going, quite literally, sideways. Just the plumbing to get the fuel to all those engines was a dreadful puzzle. Not to mention the primitive Soviet computer tech that was supposed to coordinate it all. It was like trying to fly a rocket by one of the abacuses Soviet shop clerks used in lieu of cash registers.

The N1 also triggered a breakup that's sadder than the one between Korolev and Glushko. Now racing against not only the Americans but also Chelomei, Glushko, and his own failing health, Korolev was more rushed and reckless than ever. He made a decision that several on his team, including Feoktistov, thought was insane. He declared that there was no time or budget for testing the thirty-engine monstrosity on the ground before attempting to launch. They would test the N1 in the time-honored Wrong Stuff way: just shoot it off and see what happens. Even with his history of daring stunts, this was madness. Voskresensky, who was still the most pragmatic *and* the most outspoken guy on the team, told him so. The thirty-engine rig was so ridiculously complicated that he estimated it could take three or four years of ground testing just to get it ready, and then at the very least ten test launches to see if it really would fly. To Korolev this was absolutely unacceptable.

No time, no time. They shouted about it. They screamed about it. Voskresensky threatened to quit several times. On one occasion, Korolev paced around Voskresensky, trembling with rage, shaking his fist in Voskresensky's face, threatening to beat him with a stick. Voskresensky, who was physically fading himself, laconically replied that they were a little old for such antics. Finally, toward the end of 1964, he walked off. He wasn't going to stay and contribute to what he was sure was going to be a disaster.

He never came back. A year later, December 1965, he died of a heart attack. He was fifty-two. Stalin had gotten him at last. Korolev wept at the funeral.

While all that drama was going on, OKB-1 was also developing Korolev's new generation of spacecraft called Soyuz (Union), much more sophisticated and complicated than the Vostok/Voskhod. It was three modules in a line. In front was the Orbital Module (OM), a sphere not unlike Vostok. Behind it, accessed through a hatch, was the Descent Module (DM). It had seats for three cosmonauts, still cramped but no longer squeezed in like Vienna sausages. Behind that was the Instrument Module (IM) for the engines and equipment, including unfolding solar panels for power. Soyuz would be more maneuverable than Vostok, and the OM would be able to dock in orbit. Most of that would be done remote-control from the ground or by the badly out-of-date onboard computers. The Soviets still did not trust cosmonauts with those tasks.

Leonov's one brief experience going EVA, even though it nearly killed him, was considered useful background for flying in this thing to the Moon. Where NASA would send three astronauts, two of whom would go down to the surface in the lunar lander, Korolev's obstinacy about the N1's engines meant he could only send two cosmonauts, in a modified Soyuz, with his version of a lunar lander stuck to its nose. Only one cosmonaut could land on the Moon, and the tortuous process for getting him from the Soyuz into the lander was harrowingly risky. While the Soyuz was orbiting the Moon, he would have to go EVA, haul himself to the lander module, climb in, detach, and fly down to the surface. His stay was planned to be very brief. He'd get out, plant a red hammer and sickle flag, collect a few Moon rocks, then get back in the lander, blast off, dock again with the Soyuz, go EVA again and climb back into the Soyuz, after which they'd eject the lander and head for Earth. The opportunities for deadly error were uncountable. One of the most worrisome was what would happen if a lone cosmonaut in his Michelin Man space-suit stumbled on the Moon and fell onto his back. He might find it very hard, if not impossible, to get back to his feet. *Help, I've fallen on the Moon and I can't get up!*

In October 1964, having set the mess of Moon programs in motion, Khrushchev was benched and Leonid Brezhnev took over. Brezhnev was no Khrushchev. If Khrushchev had been the class clown of Communism, Brezhnev was more its humorless headmaster. There was nothing remotely eccentric or impulsive about him. He had the stolid, saurian mid-brain of a lifelong party apparatchik. An old-school Stalinist who had gone along with Khrushchev to get along, he was not without his dissipations—another big drinker, and a chaser of secretaries—but in matters of state he was relentlessly stodgy. Government policy turned harder, more conservative and intransigent. The KGB—including a young Vladimir Putin starting in the mid-1970s—exercised uncontested power. The military was let off its leash for the brutal crushing of the Prague Spring in 1968 and the catastrophic invasion of Afghanistan in 1979.

Both of those actions were indicative of Brezhnev's relationship with the armed forces. Khrushchev had alienated them; Brezhnev embraced and indulged them. He enjoyed dressing up like a distinguished officer himself, wearing so many pounds of bogus medals on his chest that it was joked his parade tunic could start an earthquake if it fell off its hanger in his closet.

This infatuation was an immediate problem for Korolev. While Khrushchev was bloviating about his largely imaginary ICBM force, the Americans had built theirs up to massive superiority—Soviet intelligence estimated it at fifteen nukes to one. Brezhnev appreciated the space program for its propaganda benefits, but the generals had no trouble persuading him that the arms race should take precedence over the space race. He directed that they be supplied with the ICBMs, spy satellites, and space stations they wanted. Korolev's budget, which had always risen more steadily than his rockets, leveled off in 1965.

Not a single cosmonaut flew for two years after Voskhod 2. While the Soviet hare tarried and dithered through 1965 and then 1966, the NASA tortoise trudged past it. The Americans completed their entire Gemini program, ten crewed flights, in those two years, and achieved the longest flight by far to that point—two astronauts sitting in their tin can for fourteen days. They also achieved the first docking and undocking of two spacecraft in orbit, which both sides considered a crucial step toward a Moon landing. The Soviets couldn't get within kilometers of docking their crude Vostoks, more ballistics than spacecraft; it was hoped Soyuz would do better.

NASA's progress was not without some setbacks. In 1966 Gene Cernan's EVA suit bloated like Leonov's had. He sweated off ten pounds in direct sunlight before he got back into the capsule. That year two astronauts died in a fiery crash trying to land a jet in rain and snow. Their names, Elliot See and Charles Bassett, are on the Moon plaque.

By October 1965 the boredom and frustration among cosmonauts was so great that Gagarin and five others signed a long, remarkable letter to Brezhnev. "The USA has not only caught up with us, but even surpassed us in certain areas," they complained. "This lagging behind of our homeland in space exploration is especially objectionable to us cosmonauts, but it also damages the prestige of the Soviet Union and has a negative effect on the defense efforts of the countries from the socialist camp." They closed by noting that the fiftieth anniversary of the October Revolution would be in 1967. "We would like very much to achieve new big victories in space by the time of this great holiday." Along with Gagarin, Titov, Leonov, Belyayev, Nikolayev, and Bykovsky signed. It's a measure of their star status that they dared to. They gave it to Kamanin, who promised to pass it up the spout. It's doubtful that it ever made it to Brezhnev's desk.

In December 1965, the end of a frustrating year, the month of Voskresensky's death, OKB-1 threw a morale-building party in its canteen at Kaliningrad for five hundred of the staff and a handful of cosmonauts, including Gagarin and Leonov. Leonov recalled:

> A jazz band, made up of engineers and others from the enterprise, played. We danced. Sergei Pavlovich [Korolev] danced with his wife, Nina Ivanovna, and with several other women. He was in great demand that evening and he

liked to dance. He was very sociable, not at all the person
he was when he was at work. He rarely had the opportunity
to attend such parties because of his hectic schedule and
heavy workload. But among those he trusted and loved he
became almost a different person. He was relaxed. He told
jokes. He was the life and soul of the party. [He] seemed
full of vitality that night. He had the physique of an ox, and
looked strong and stocky.

In fact he was worn out. Because of his irregular heart he periodically spent a week or more in the hospital. There were times when his
blood pressure dropped to worrying lows and the slightest activity
exhausted him. That month alone he'd spent four days at the Kremlin Hospital, a private clinic that provided the country's top-tier elite
with the best care available in the Soviet Union. That was a long stay
for what was called a "routine examination," and all that turned up
was some small polyps in his intestine. He was scheduled to have
them removed in January, after his fifty-ninth birthday.

On January 10 he and his wife had family and friends for dinner
at their home, a charming wood-frame house on an acre of trees and
shrubs in a Moscow suburb. The government had given it to him in
1959. In America it'd be a middle-class family home. In the USSR
it was a palace, with its gleaming woodwork and large fireplace. In
his study on the second floor Korolev sat under his portrait of Tsiolkovsky. On this night his servants—servants! in the workers' paradise!—laid out some meat and cabbage *pirozhkis*, and he produced a
bottle of fine Armenian cognac, which Churchill had been very fond
of. As Khrushchev had known, even in a Communist police state

it's good to be the (anonymous) king. This was the night when he sat up with two favorite cosmonauts, Gagarin and Leonov, telling them about his ordeal under Stalin. Leaving at 4:00 a.m., Gagarin and Leonov were in shock. They'd never heard any of this. "We began to realize there was something wrong with our country," Leonov later wrote with masterful understatement.

On January 12 Korolev quietly celebrated his birthday. Two days later he was back in the hospital to have those polyps removed. It was supposed to be a simple and quick procedure. In *This New Ocean*, William Burrows writes that Dr. Boris Petrovsky, the minister of health, knew that the Chief Designer was an important man with important friends. "He therefore decided to impress his celebrated patient with his own knowledge—in the process neglecting to get a second opinion—and take it upon himself to perform the operation." This despite his being more of a bureaucrat now than the surgeon he once was, more familiar with pushing paper than pinching polyps. With his ruined jaw Korolev could not be intubated—thank you, Comrade Stalin—so he lay on the table with an oxygen mask on. Petrovsky was soon in trouble. "Plucking out the polyps ruptured a blood vessel that caused severe hemorrhaging," Burrows relates. Petrovsky made an incision to try to stanch the bleeding, and was shocked to see "a large malignant and rapidly spreading" abdominal tumor that had gone undetected during Korolev's *four days* of medical exams the month before. It was said to be the size of "two fists." Panicking, Petrovsky spent the next seven hours hacking away at Korolev's guts, trying to dig the tumor out. As Korolev began to have heart spasms, Petrovsky finally sent for help. The best surgeon on staff "was reached at a nearby resort and eventually rushed into the operating room and up to the gutted patient. He took one look at Korolev and announced that he did not operate 'on dead men.' Then he stalked out of the room."

Korolev died soon after, a victim of Wrong Stuff medicine. Coordinating Leonov's spectacular, nearly deadly space walk had been his last hurrah. His obituary in *Pravda* was the first time he was identified in print as the fabled Chief Designer. Boris Chertok was tasked with writing it. He included the line, "Korolev remained an ardent patriot and steadfastly pursued his goal—to fulfill the dream of spaceflight, despite years of unjust persecution." When his editor saw the draft, he crossed that sentence out. With the unregenerate Stalinist Brezhnev in power even an oblique reference to Stalin's "unjust" reign was unprintable.

Kamanin wrote that the death fell like "an avalanche" on the whole space community, and they all seemed to be at the lavish funeral in the ornate House of the Unions, a very high honor; Lenin and Stalin were both laid out there. "Kaliningrad" would later be renamed Korolev as another honor. Long lines snaked through the hall, ordinary citizens mourning a man whose name they hadn't known until now. Gagarin was the last speaker before the urn carrying the ashes was placed in its niche in the Kremlin Wall. Privately, he was enraged and cursed the minister of health for conducting a procedure he should have let a more competent surgeon do.

Back in 1964, a brooding Korolev had said to Chertok, "We absolutely must not let the Americans make the first soft landing," speaking about unmanned lunar probes. Both sides considered probes necessary to check conditions on the Moon before manned landings.

He tried and tried. Luna 4 never made it past Earth orbit. Lunas 5 and 7 crashed into the Moon. Luna 6 flew right past it, missing it by no less than 161,000 kilometers. The last launch Korolev supervised, Luna 8, also crashed, in December. He was looking forward to trying again with Luna 9 when he died. Seventeen days after his passing, Luna 9 was launched. In February 1966 it was the first Earth vehicle in history to make a soft landing on the Moon, a posthumous victory.

It's been said that Korolev's passing marks the end, or at least the beginning of the end, of the race for the Moon. Even though in his last years, especially in his bickering with Glushko, he did as much to impede the program as advance it, he still left a void. He had been the space program's driving, driven virtuoso since the postwar years. No one expected his successor, Vasily Mishin, to fill his shoes. Mishin had been Korolev's devoted assistant from the start, but he wasn't the visionary Korolev was, nor the megalomaniac. Where Korolev was more of a manager than an engineer, Mishin was the opposite. He didn't inspire or bully his staff the way Korolev had, he couldn't cajole the kind of Kremlin support Korolev had, and he couldn't coordinate all the other agencies and phenomenally unpredictable supply lines like Korolev had. Right away, Mishin put his mark on things by plotting a routine of methodical planning and testing before risking cosmonaut lives. He planned for a series of unmanned Soyuzes code-named Zond (Probe) to go around the Moon first. If that went well, he wanted to send three two-man teams also on circumlunar missions. If all went well again, one of those teams would then attempt a landing. Alexei Leonov and a new partner, a civilian engineer named Oleg Makarov, were one of the teams. Compared to the elaborate course of training Apollo teams were going through, Leonov's sounds like a peculiar Soviet combination of the lackadaisical and the life-threatening. It didn't amount to much more than practice landing a helicopter with its engine shut off, thought to approximate what falling to the surface in a lunar lander would be like.

By early 1967, the Kremlin was hustling Mishin to move things along. It was a big year in Soviet history, the fiftieth anniversary of the October Revolution, and Brezhnev wanted some fireworks from the space program. He got them, but not in the way he expected.*

---

* A postscript: David Scott and the other two Apollo 15 astronauts soon found themselves testifying before a closed-door Senate hearing about carrying other objects to the Moon besides the plaque and the little figurine. They also took stamped envelopes, which a dealer sold for them to memorabilia collectors on their return to Earth. Astronauts cashing in on their trip to the Moon was hugely embarrassing to NASA, but because public support for Apollo missions was already very soft, a lid was kept on the scandal. Still, those three astronauts never flew again. Meanwhile, the artist Van Hoeydonck reportedly hoped that having the first sculpture on the Moon would make him "bigger than Picasso." Ever hear of him?

12
THIS DEVIL SHIP!

Truly, there never was a time when we worked in
peace, without being hurried or pressured from
above. The unskilled, totally bewildered, high-ranking
bureaucrats believe that they are fulfilling their duties
if they are shouting "Let's go, let's go!" at people
who don't even have time to wipe the sweat off
their brows.

—VASILY MISHIN

In April 1967, Vladimir Komarov had the honor of riding in the
brand-new Soyuz 1 on its maiden flight.

It started falling apart on the first orbit.

"This devil ship!" he reputedly radioed as one malfunction after
another cropped up. "Nothing I lay my hands on works properly."

His trip was planned as a long and busy one, but ground control decided to bring him down as soon as possible. "As he began his descent into the atmosphere," Doran and Bizony write in *Starman*, "Komarov knew he was in terrible trouble. [American] radio outposts in Turkey intercepted his cries of rage and frustration as he plunged to his death, cursing forever the people who had put him inside a botched spaceship. . . ."

The story of Komarov's last mission vies with Bondarenko's and Nedelin's demises as the most gruesome of all the tales from the Cosmodrome. There are differing versions, from the lurid to the levelheaded, and serious disagreements among the tellers. But one thing cannot be disputed: Vladimir Komarov managed to survive the ridiculous Voskhod 1 mission of 1964 only to die a horrible death in the just as slapdash Soyuz 1 mission in 1967. If he didn't plunge to his death screaming in "rage and frustration," he should have.

The celebrations planned for the fiftieth anniversary of the revolution would kick off with Lenin's birthday in April and then May Day, the annual celebration of international Communism. The Kremlin had let Mishin know that it would dearly love to celebrate a triumphant space mission on one of those days. The horrible deaths of the three American astronauts Grissom, White, and Chaffee in the Apollo 1 fire that January surely meant there'd be no competition from NASA for a while.

Given that no cosmonauts had flown for two years, the mission planned for finally getting the manned program back in motion was ambitious. Soyuz 1 would launch with just Komarov onboard. Soyuz

2 would launch the next day carrying three cosmonauts. They would rendezvous and dock in orbit. Then two cosmonauts would exit Soyuz 2, spacewalk over to Soyuz 1, and join Komarov. The vehicles would undock and fly home. Docking, undocking, and transferring crew would all be necessary steps for both a Moon landing and working with space stations, the latter being of great interest to the Soviet military.

There was just one hitch: Soyuz was far from ready to fly in the spring of 1967. Mishin had fallen well behind in the confusion after Korolev's death. Kamanin called the situation "criminally chaotic." Three unmanned test flights from Baikonur had all just ended in catastrophic failure. In November 1966, the first Soyuz (cover name Cosmos 133) was launched. A second one was to go up the next day, then they'd be maneuvered and remotely docked. In orbit Cosmos 133 went into a roll that Chertok and the others couldn't correct. The second launch was scrubbed, and Cosmos 133 made an uncontrolled reentry and blew up. Two weeks later, the rocket carrying the next Soyuz blew up on the pad, killing three pad workers. Then in February 1967 a third Soyuz actually made it into orbit. Everything went well until its descent, when its heat shield melted through. It landed well off course in the Aral Sea, half as hot as a chunk of the sun, burning a hole through the ice and sinking to the bottom. Chertok remembered an engineer joking it was "so ashamed of its inaccurate landing that it drowned itself."

There were other grim warnings. Engineers listed more than one hundred "anomalies" in telemetry, communications, and guidance systems that should be addressed before a man flew in a Soyuz. Even for the Soviets it would be unprecedentedly reckless to send up four men in vehicles that had that many problems and hadn't passed even a single unmanned test. As time grew short, a harried Mishin telephoned Brezhnev to express his anxieties about the mission. He got the standard Kremlin response: The show must go on. Make it so. Kamanin wrote in his journal: "I am personally not fully confident that the whole program of flight will be completed successfully," but then added, "although there are no sufficiently weighty grounds to object to the launch." Really?

According to Doran and Bizony, Komarov was as worried about the mission as anyone. Venyamin Russayev, Gagarin's personal KGB escort and friend, told them, "Komarov invited me and my wife to visit his family. Afterwards, as he was seeing us off, he said straight out, 'I'm not going to make it back from this flight.'"

Gagarin played a curious role. In 1964 he had been grounded from flying any aircraft, not just a spacecraft. The Kremlin thought he was far too valuable as an icon to risk losing, and given his drinking and carousing they preferred that he didn't even drive, let alone fly. He accepted this obediently if reluctantly for a while. But with Korolev's death in January 1966 something snapped in him. He declared his intention to be the first man on the Moon and bring Korolev's ashes with him. He clashed with some fellow cosmonauts, who didn't want to be bumped from missions to make room for the pouty celebrity. He'd had his heroic adventure; they wanted theirs. He badgered Kamanin and Mishin until they reluctantly caved in and gave him an assignment: Komarov's backup. Though he went through all the training, it's extremely unlikely the bosses would have let him go. It was just busywork to placate him.

At 1:00 a.m. on April 23—they missed Lenin's birthday by one hour—Komarov arrived at the launchpad. He wasn't wearing a spacesuit for the flight, just the wool outfit. As his backup, Gagarin rode the elevator up to the capsule with him. Everyone at the Cosmodrome held their breath as Komarov lifted off a couple hours later. As for why they let him fly in a vehicle they knew to be deeply flawed, we should remember that these may have been good people but they'd never lived a day in a democracy, a place where they were free to speak their minds. They'd spent all their lives, as had their ancestors, in a totalitarian Big Man state where you did not disagree with the Big Man, or did it very, very gingerly, and at no small peril.

Komarov had barely made orbit when the problems started. One of two solar panels failed to deploy, which meant severely reduced electrical power. The system that guided automatic maneuvering of the vessel failed, which meant that docking would not be possible, so the Soyuz 2 launch was canceled. Then attitude control failed, and the ship started slowly tumbling end over end. Now Chertok and the rest at mission control frantically scrambled to figure out how to get Komarov home safely. An attempt to begin reentry on orbit seventeen, about twenty-four hours into his flight, failed. Then again on the next orbit. On orbit nineteen they went fully manual, which the Soviets did only in dire emergencies. Komarov executed a very complicated series of activities he'd never trained for—Chertok and the team had come up with them on the fly—and finally began his reentry. His communications cut off during the descent, which was normal. But he was not heard from again. His parachutes failed, and he made what's known in the profession as a "ballistic reentry." A recovery helicopter spotted his smashed, blackened DM lying on its side in a field near the Russian city of Orenburg, just over the border with Kazakhstan. The ragged parachutes were stretched out beside it.

As the chopper approached, the small rockets that were supposed to soften the landing belatedly ignited, lighting up the whole DM in a terrifically intense fire. By the time the team landed and approached, the DM had actually melted. When the fire was finally out, what was left of Komarov was not recognizably human. He was a twisted lump of what looked like burnt chewing gum. They had difficulty separating him from the molten ship.

Was Komarov screaming and cursing as he hurtled to his death? That version comes from a 1972 issue of the magazine *Ramparts* that ran an interview with a National Security Agency whistleblower named Perry Fellwock, using the cover name Winslow Peck. It's known as the first time the NSA's size and activities were revealed in print. Fellwock said that a listening post in Turkey intercepted Komarov's radio communications with the ground.

> They knew what the problem was for about two hours before he died, and were fighting to correct it. It was all in Russian, of course, but we taped it and listened to it a couple of times afterward. Kosygin [Brezhnev's number two] called him personally. They had a video-phone conversation. Kosygin was crying. He told him he was a hero and that he made a great achievement in Russian history, and that they were proud and he'd be remembered. The guy's wife got on too. They talked for a while. He told her how to handle their affairs, and what to do with the kids. It was pretty awful. Toward the last few minutes, he began falling apart, saying, "I don't want to die, you've got to do something." Then there was just a scream as he died. I guess he was incinerated.

In 2018 space historian Asif Siddiqi obtained a copy of what he believed was the official record of Komarov's flight, including a transcript of his communications with ground control. It was found among Kamanin's papers and, like so much other Soviet memorabilia, sold at auction. Siddiqi used it as the basis for a slim book, *Soyuz 1*, in which he categorically refutes Fellwock's version of events. Although he admits that Komarov shouted at ground control early in the ordeal, he believes the transcript ultimately shows a cool professional going through all the routines to get himself home. The transcript ends with Komarov saying, "I feel excellent, everything is in order. . . . Thank you to everyone."

Komarov was an experienced fighter pilot, and as a cosmonaut he had already survived one ridiculously slapped-together space vehicle before this one. If anyone had Tom Wolfe's Right Stuff, it was him. In the last moments of his fiery plunge to the ground he knew he was about to die when he felt no jolt of sudden deceleration, meaning the parachutes had not opened. Maybe he really was calm and stoic about that. But it does raise the question: If we don't believe the version told by the NSA spook, why do we trust a purported Soviet transcript?

The Kremlin had hyped the intended Soyuz-Soyuz docking mission and could not hide the tragic outcome. Although the government released scant details, it was still the first time the Soviet public had heard of a deadly accident in the space program. Komarov's charred and mangled corpse was flown to Moscow. There was no question of an open-casket funeral, but the air force brass wanted a photo for documentation. The casket lid was lifted, Kamanin wrote in his diary, and "on white satin lay what was only recently the cosmonaut Komarov, and now it has become a shapeless black lump." The astonishing, appalling photo is online.

Gagarin and Leonov were among those present. They told Kamanin they blamed Mishin's terrible mismanagement for their friend's death. Kamanin did not disagree, and added in his diary the opinion that Mishin was a drunk. In Mishin's defense, it's easy to see how anyone in his position would be driven to drink.

Komarov's tragic death shocked and saddened the Soviet people, who for a decade had become used to hearing only about triumphs in space. But the real trauma was yet to come.

The Soyuz 1 debacle put the Soviet manned space program on pause again, and definitively ended any hope Gagarin had of flying to the Moon or anywhere else. He was shoved back behind a desk, a cosmonaut-bureaucrat. Despondent, drinking, he was such a sad sack about it that the bosses relented yet again and let him resume MiG training in 1968.

The Moscow area was in winter's lingering grip on the gloomy morning of March 27, with rain and sleet falling from low clouds onto deep snow. Not a great day for flying. In addition, for Gagarin the day started with a series of bad omens. On his way to Chkalovsky air base near Star City his car broke down and he had to grab a bus. "Then he realized he had forgotten his gate pass, so he had to return home to pick it up before returning to the Chkalov airfield," Burgess and Hall write in *The First Soviet Cosmonaut Team*. At the base, "his superstitious colleagues chided him about having a bad-luck day, and that he should reconsider flying." He waved them off. At

10:19 a.m. he took off in a two-seater MiG-15 trainer seated in front of his instructor, Colonel Vladimir Seryogin, the war hero and trainer of Tereshkova. The MiG disappeared into the low clouds. It was only 10:31 when they called in that they were cutting the flight short and returning to base. They gave no reason.

They were not heard from again.

Gagarin's good friend Leonov was nearby on the morning of March 27, trying to conduct parachute practice for a new crop of cosmonaut hopefuls. Because of the filthy weather he had just cut the session short when he heard two booms in the distance. It sounded to him like a jet breaking the sound barrier, followed by an explosion.

At Chkalovsky he was informed that Gagarin and Seryogin had gone silent and were forty minutes overdue. A search pilot saw smoke rising from a patch of birch forest. He landed and struggled two hours through waist-high snow to reach the site. The MiG had smashed into the ground at high speed, punching a crater through the snow and scattering a wide field of debris. The spot was sixty-five kilometers from the airfield. Only small bits of Gagarin and Seryogin could be found among the strewn wreckage. "A few days before, I had accompanied Yuri to the barber to have his hair cut," Leonov recalled. "I had stood behind Yuri talking while the barber worked. When he came to trim the hairs at the base of Yuri's neck he noticed a large, dark brown mole. 'Be careful not to nick that,' I told him." Leonov now recognized the mole in a small piece of flesh retrieved at the crash. "'You can stop searching,'" I told the rescue workers. 'It's him.'" Hundreds of soldiers combed the forest for debris. Curiously, there was no sign of the pilots' two parachute packs at first. Three days later, "KGB agents located the missing packs which had been hidden beneath a pile of manure in a neighboring village,"

Burgess and Hall relate. Looters had probably intended to sell the silk, "totally unaware of the identity of one of the pilots." They did find Gagarin's wallet. There was a photo of Korolev in it.

Soviet society was paralyzed by grief at the news that the thirty-four-year-old was dead. A day of national mourning was declared—the first ever for a Soviet citizen who was not a head of state. The Gagarin cult, which had already been as fervid and wide-spread as any since Stalin's, now went transcendental and mystical, like Elvis's after his death. Statues were erected all over the vast-ness of the Soviet Union, including an immense one in Moscow in which he seemed to be blasting off to the stars under his own power. The city of Gzhatsk was renamed Gagarin. A tiny shack said to be the home where Yuri was born, transported to the city from Klushino and venerated like a sacred shrine, stands on Gagarin Street. Star City became home of the Yuri Gagarin Cosmonaut Training Center.

Leonov and Titov were assigned to the official government com-mittee investigating the crash. Their efforts generated some thirty volumes of findings and conclusions—which, typically, the govern-ment kept secret. But the conclusions were inconclusive anyway. Leonov was persuaded by the theory that a new supersonic Sukhoi Su-15 fighter, bigger and more powerful than the old-fashioned MiG trainer, veered so close to the little jet due to the poor visibility that its turbulence sent them spinning into the ground. Another possi-bility was that the MiG had swerved to avoid a weather balloon. Nei-ther could be confirmed.

As usual, the government's secretiveness provided rich soil for rumors and conspiracy theories. Gagarin was known to have got-ten drunk at a party a few nights earlier, and it was rumored he was drunk again when he took off in the MiG. Leonov was outraged by that one, though Gagarin's behavior over the years suggested it was not totally implausible. Another rumor was that Gagarin and Seryogin had been shooting at wild deer from the cockpit and lost control of the jet, which would only begin to approach plausibility if one of them had LBJ sitting on his lap. A popular theory insisted that Brezhnev had the MiG shot down to rid himself of Khrushchev's troublemaking pet. Another said that the CIA assassinated Gagarin to get rid of a Soviet hero. There was the one that, like Elvis, Gagarin had not really died when they said he did; he was in a sanatorium, and would return someday. And the inevitable too-close encounter with a UFO, the inhabitants of which shot him with a "brainfreeze ray."

With Gagarin gone, there was absolutely no way the bosses were going to let Gherman Titov ever risk another spaceflight. They banned him from flying jets as well. Bored, he would resign from the corps in 1970.

Six months after Gagarin's death, in September 1968, CIA listening posts heard a trio of cosmonauts' voices reciting telemetry as their capsule went in "for a landing on the surface." Shock waves raced through NASA. Had the Soviets sneakily beaten Apollo to the Moon?

No, they had not. But they did have a vehicle flying around the Moon, Zond 5. Zond was another example of the Soviets giving an existing spacecraft a new name and pretending it was . . . new. The original Zond craft were interplanetary probes aimed at Venus and Mars. Then Mishin started using the name for attempts to orbit the

Moon using unmanned Soyuz ships. He launched Zond 4 in March 1968; not a lunar approach, but instead a test shot well out into space and back. Things went well until its reentry trajectory went off and the controllers activated its self-destruct mechanism rather than have it fall outside the Soviet Union and into "enemy" (Chinese) hands.

Zond 5 was doing its pass around the Moon on September 18 when Western listening posts picked up the jarring transmissions, apparently from its three-cosmonaut crew of Valery Bykovsky, who'd flown Vostok 5 in tandem with Tereshkova's Vostok 6; Pavel Popovich, who'd been one of the Heavenly Twins; and Vitaly Sevastyanov. They were *not* aboard Zond 5. They were actually at the mission command center in Crimea. Thinking it would be a fine prank on their competition, they asked the techies there to bounce a transmission up to Zond 5 and back. Popovich later recalled, "I took the mic and said: 'The flight is proceeding according to plan. We're approaching the surface.'" The others recited telemetry data. It's not known how they managed to keep straight faces. It was a good month before the Americans realized they'd been punked.

Zond 5's actual crew was two turtles and Ivan the dummy. The probe made it back to Earth, but like Zond 4 it overshot its planned landing site and splashed down in the Indian Ocean. American and Soviet navy ships raced each other to it. The Soviet ships got there first and retrieved the capsule as the Americans watched from a distance. The turtles were unharmed by their adventure in space. It was the scientists on Earth they had to worry about. Like their predecessors, the turtles were soon euthanized and dissected for study.

Zond 6 flew the same circumlunar route that November and took some great pictures of the Moon on its way around. But various of its systems began malfunctioning on the way back, its parachutes failed on reentry, and it crashed in Kazakhstan. This time its turtle crew did not survive.

Zond 5 and 6 spooked the Americans into wondering if the Soviets really were close to a manned shot. In fact, Leonov and Makarov asked to go right away, but Kamanin and Mishin told them that was impossible. NASA sped up its Apollo timetable. Originally Apollo 8, scheduled for launch in December, was just going to orbit Earth. That plan was now changed for a much ballsier, dare one say Soviet-style, leap: Apollo 8, launched on December 21, 1968, was the first manned spacecraft to orbit the Moon and return safely.

Now it was the Soviets who were spooked. They knew that the Americans were just a few steps away from putting boots on the Moon. Mishin launched the first N1 in February 1969. The behemoth "shuddered and began to lift off," Chertok recalled. "The roar penetrated underground through several meters of concrete." But a little over a minute into its climb, that complicated thirty-engine booster developed problems and shut down. "At a slight angle to the horizon, [the rocket] was still moving upward; then it tilted over, and, leaving a smoky trail, without breaking up, it began to fall." It smashed into the desert and exploded some thirty miles out from the Cosmodrome. "It isn't alarm and it isn't dismay," Chertok wrote, but "more a certain complex mixture of intense inner pain and a feeling of absolute powerlessness that you experience while watching a crashing rocket approach the ground."

March 1 was Chertok's fifty-seventh birthday. His colleagues crowded into his office with a bottle of cognac and a guitar, and sang him a song they wrote. The lyrics are translated in his book:

I know, my dear friends, many years will pass by,
And the world will forget all our pains,
But in the wreckage of many a rocket
The mark we made always remains.
Let us be stumbling drunk tomorrow,
The rocket flew away; pour another glass . . .

Then came the watershed month in the Moon race, July 1969. Mishin tried to launch his second N1 that month—on the Fourth of July, as it happened. The giant rocket struggled to only three hundred feet off the pad, then crashed back, exploding in an enormous orange mushroom cloud of burning fuel that lit up the night (like fireworks?) and tossed rocket parts in a ten-kilometer radius. Luckily, since Nedelin's day launch safety precautions had been improved and no one was standing around to die horribly this time. It would take a year and a half to repair the pad.

Mishin launched Luna 15 from Baikonur on July 13. Apollo 11 took off from Cape Canaveral three days later. When the Americans tracked Luna 15 on its way to the Moon they demanded assurances from the Kremlin that it wasn't some sort of attack ship. The Soviets, in a rare moment of honesty, assured them it was not. Both vessels were orbiting the Moon by July 19. The Americans touched down on July 20. Incredibly, they could see, and film, Luna 15 as it passed over their heads. As if Apollo 11's triumph wasn't galling enough for the Soviets, the probe suffered a signal failure as it was descending and crashed into one of the Moon's mountains on July 21.

The race to the Moon was truly over. Cosmonauts would stop training for it in 1970, as the Soviet space program was refocused on the space stations the military wanted. From then on Kremlin propaganda firmly denied that the Soviet Union had ever intended to put men on the Moon. As of this writing, no Russian has set foot there (that we know of).

Mishin would try two more times to get an N1 to fly, in 1971 and '72. Both failed. Chertok would direct the last launch. Once again he would watch the mammoth rocket lift off successfully, only to detonate in an apocalyptic fireball.

In America, Apollo 11's achievement had a curious effect on the public psyche. A brief period of euphoria, as though their team had beaten a rival team in an interplanetary World Series, was followed by a general deflation of interest. Within two years, when Apollo 14's Alan Shepard was videoed shanking golf balls across the Moon's Fra Mauro formation—likely the most serious gaffe in the always propaganda-challenged NASA's history—the disinterest turned to outright hostility. It's rarely mentioned in books, TV shows, or movies about the space race, which tend to trade in shiny-eyed nostalgia and patriotic rah-rah, but in fact a majority of Americans had always questioned why they were spending so much money and effort on beating the Soviets to the Moon. There were so many pressing concerns on the ground, like poverty, hunger, racial injustice, race riots, political assassinations, the Vietnam War, and the nuclear arms race. In opinion poll after poll through the 1960s, up to 60 percent of respondents said the government was spending too much on it. The conservative commentator William F. Buckley called it "The Moon *and* Bust." Sociologist Amitai Etzioni wrote a book called *The Moon-doggle.* And most memorably of all, Gil Scott-Heron wrote the poem "Whitey on the Moon," which he rapped to Conga drums. It's about being black and struggling to survive in the ghetto while "Whitey" is puttering around on the lunar surface.

Whitey wasn't there long. Apollo 17 in 1972 would be the last one. It would leave the Moon's surface just two weeks after the last N1 failure. From then on the Americans, like the Soviets, would be directing their programs to near-Earth orbit, the realm of satellites and Soyuz, shuttles and space stations.

13

RED MAN
FALLING

B oris Volynov was the first Jew in space. He came very close to
being the first Jew to die there.

Although one of the original cosmonaut corps, Volynov somehow kept missing out on going to space. It finally looked like he was
going to get his shot in 1966, flying in Voskhod 3 with Georgi Shonin. Then Korolev died just ten days before their launch date, and
Mishin killed the Voskhod program to move on to Soyuz. Shonin
later said he was relieved. He and Volynov were to spend twenty days
in orbit, to beat the Americans' record of fourteen days. Shonin said
he couldn't imagine how two men stuffed into the inhumanly tight
space of a Vostok-er-Voskhod capsule for three weeks could have
survived it.

After Komarov's ghastly death, Mishin's team spent more than a year trying to solve the devil ship's myriad problems, along with everything else he was trying to make happen—the N1, the Lunas, the Zonds. In October 1968 they felt ready to launch a cosmonaut in Soyuz 3. He successfully completed eighty-one orbits but failed to dock with the unmanned Soyuz 2. In January 1969 it was Volynov's turn to try, in the tandem Soyuz 4 and 5 mission, which was them taking a mulligan for Komarov's intended mission. In orbit, Soyuz 4, with one cosmonaut aboard, and Soyuz 5, carrying three, successfully docked. Then two of the guys from Soyuz 5 went EVA, which was much easier to do with the compartmentalized Soyuz than it had been for Leonov in the Voskhod. Now the pilot could stay in the pressurized DM while the space-walkers in their spacesuits depressurized the OM and climbed out. From there they worked their way across to Soyuz 4, entered that depressurized OM, and repressurized it.

The two crafts then headed home—Soyuz 4 now with a crew of three, while Volynov was the lone pilot in Soyuz 5. He was wearing his wool leisure suit, strapped in tight inside the DM when the retro-rockets fired to slow his speed for de-orbiting. Then explosive bolts jettisoned the Vostok-like OM from in front of the DM. Next, more explosives were supposed to detach the IM that was behind him, attached to the heat shield. That's when things went seriously tits up. With the IM detached, the DM would flip around to descend heat shield–first. But the IM did not detach. Basically it was the same problem that had been plaguing the program since Gagarin's flight *eight years before*, and it still wasn't fixed. Because the IM was still attached, the module did not flip around like it was supposed to; it shot into the atmosphere face-first, the wrong way around. This was catastrophic. With no shielding, the DM would surely burn up, cooking Volynov inside it. "I looked out of the window of the capsule and saw the flames," he said later. "I said to myself, 'This is it. In a few minutes I'm going to die.'"

He reported his dilemma to ground control, hoping against hope that they'd have a solution. They did not. Instead, one of the military men in the control center took off his cap, tossed a few rubles in it, and began passing it around the room. They were taking up a collection for Volynov's family, since they agreed with him that he was surely a dead man falling.

As the DM with the IM still attached came screaming into the atmosphere, the unprotected nose began to burn at five thousand degrees Fahrenheit. A cone of superheated plasma surrounded the ship and began to melt it. Volynov, in his leisure suit, hanging face down in his seat, felt the heat building up and could smell burning rubber. That was the gasket for the front hatch melting away. He knew that when it went he'd light up like a human torch. Just to make the trip really exciting, the fuel tanks in the IM behind him started blowing up.

Then he had a stroke of luck, the same that had saved the lives of Gagarin and Titov. The hot plasma burned through the struts, keeping the IM attached. It broke away and vaporized as it fell. The DM flipped around. Now it was falling the way it should, heat shield first, Volynov with his back to the ground. The small drogue parachute in the upward-pointing nose of the ship opened, then the larger main chute. Then the lines for the two got tangled. The DM hit the ground much faster and harder than it should have. The impact was so hard that Volynov was shot straight through his safety harness and flung face-first into the instrument panel, breaking his jaw and knocking out a few teeth. But he was alive.

Because of all the problems, he had crashed down well off target—standard operating procedure for Soviet spacemen—on a slope in the frigid Ural mountains. His blackened ship hissed and crackled in the snow. The wind howled around it and flapped his twisted parachutes. The temperature outside was -40 degrees Fahrenheit.

There are two versions of what happened next. In one, Volynov simply sat in his battered capsule, shivering in his leisure suit as the temperature inside gradually lowered to match the way-sub-zero outside, hoping the rescue helicopters found him before he froze to death. They did. He asked them for a smoke and was sad to hear they only had Shipkas, among the worst of bad Soviet cigs.

The other version is more fun and thus, naturally, more popular. In this version, the helicopters spot his charred capsule smoking in the snow. They drop a rescue crew who, gasping for breath in the frozen air, trudge over to the capsule—and are stunned to find it empty. Where the hell is the cosmonaut? Did God pluck him out of the capsule at the last minute? Or the ghost of Lenin? Looking around, they see splashes of blood in the snow. Was he attacked and dragged away by wolves, or bears? But no, the only tracks they see are a man's footprints leading away from the capsule, trailing occasional spots of blood. As they follow them they see a thin wisp of smoke up ahead. It comes from the chimney of a little shack. They find Volynov inside, wrapped in blankets by the fire, being looked after by simple, good-hearted peasants. He glimpsed their cabin just before the Soyuz smashed into the snow, and decided to try to reach it rather than freeze to death or be ripped apart by bears.

Clearly that's the more entertaining and less plausible story. But once again, because the Soviets covered up and lied about it all, claiming that Volynov landed as planned after an uneventful descent, they left lots of room for alternative facts.

Volynov still wasn't done skirting death. Four days later Moscow feted the Soyuz 4 and 5 cosmonauts with the usual motorcade and state dinner. As the motorcade reached the Kremlin's Borovitsky Gate, a young army deserter in the crowd lining the road unloaded two Makarov pistols into what he thought was Leonid Brezhnev's limo. The limo was full of cosmonauts. Their driver was killed. In Stalin's day the shooter would have been convicted at a show trial and executed, and possibly his family and friends as well. Brezhnev decided to handle it more quietly. The young lieutenant was declared insane and kept in a mental institution until 1990, sort of the Soviet John Hinckley Jr.

And Volynov *still* wasn't done cheating death. In 1976 he would go back into space in another Soyuz, to rendezvous with the space station Salyut (Salute). If Soyuz was the devil ship, Salyut was the answer to the age-old questions: Is hell real? If so, where is it? The answers were yes, and it's in orbit.

Salyut's roots went back, ironically, to CORONA. The American spy satellite was very successful at snapping photos of sites the Soviets really wanted to keep secret: the Baikonur and Plesetsk cosmodromes, military airfields, naval ports and submarine pens, etc. But as good as CORONA was, it did have some drawbacks besides the complicated midair retrieval routine. For instance, unlike a U-2 pilot, a CORONA satellite couldn't compensate for cloud cover over Site A by changing course to fly over Site B. So although CORONA took many excellent, highly useful photographs of important Soviet facilities, it also took many excellent, totally useless photos of clouds over important Soviet facilities. This got some American military intelligence types thinking about how to reintroduce the human factor—put actual eyes in the sky.

The air force came up with the Manned Orbiting Laboratory (MOL, pronounced "Mole"). It was called a laboratory to give it a peaceful, scientific cover, but it was purely a military project. Starting in 1964, fighter pilots were secretly trained at Edwards Air Force Base's ARPS (Aerospace Research Pilot School), run by none other than Chuck Yeager, as both astronauts and spies. The plan was for them to fly into orbit in a modified Gemini capsule pre-attached to a small space station, which was basically a super-high-resolution camera with some cramped living and working space around it. They'd spend a month up there taking hi-res photos of Soviet sites, then fly the Gemini capsule back to Earth.

When Soviet spies in the US got wind of this program and reported it, the Soviet military weren't fooled for a second by the "laboratory" bluff. They asked Chelomei to develop their own top secret counter-MOL, called Almaz (Diamond). According to space historian Brian Harvey in *The New Russian Space Programme*, his design included "crew quarters, radar remote-sensing equipment, cameras, two small re-entry capsules for sending data back to Earth and even . . . rapid-fire cannon to defend the station against attacking American spaceplanes!"

MOL's astro-soldiers were all trained and the program was proceeding toward deployment when Richard Nixon took office in 1969 and ordered the Pentagon to cut back on expensive projects of dubious value. MOL was one of the programs terminated. So Almaz was another first for the Soviets, the world's first space station, when it was put into orbit in April 1971. Like the Americans, the Soviets played it as a civilian science project. It had the cover name of Zarya (Sunrise) right up until launch day, when Mishin suddenly realized that Zarya was already being used as the code name for the mission control center. It would have made communications a bit confusing. "Zarya, this is Zarya. Come in, Zarya." So he hastily changed it to Salyut (Salute), because April 1971 was the tenth anniversary of Gagarin's flight and they were "saluting" him. They had aimed for liftoff on April 12 to make it a precise match, but precise matches were not a strong suit of the Soviet space program and it took off a week late. When it did, it still had ZARYA painted on it in bright red Cyrillic letters because there'd been no time to paint over it. If that doesn't say all you need to hear about how ad hoc the Soviet space program was, wait, there's more.

Whatever they wanted to call it in public, Salyut was military hardware to be crewed by military cosmonauts and stuffed with military surveillance gear. It was a far, far cry from the giant, majestically rotating, *2001*-style space stations Korolev and von Braun had dreamed of building. It was no International Space Station (ISS) or Mir. It was not even as elaborate as it had been on Chelomei's drawing board. It was just a tube about sixty-five feet from end to end, with a docking station at one end, life support and other equipment at the other, and a cramped, amenities-free living/working area in the middle. It was an orbiting lead pipe.

Mishin was not a fan of this new focus on space stations. As he continued to try making Korolev's colossally flawed N1 viable, he proposed a bold new plan to plant a Soviet colony on the Moon by the end of the decade. The Kremlin turned him down flat. Mishin then neglected the cosmonauts' space station training. Part of the reason was more of the personal animus that tripped the Soviets up all the time: The space station was Chelomei's baby, and he and Mishin never got along. Salyut was also purely military despite the usual pretenses, and Mishin resented how much the military were involved in the space program, believing that it should be more of a civilian, scientific enterprise.

As a result, the Soyuz/Salyut crews were not as well prepared as they should have been. It showed. The first three-man team sent to Salyut in Soyuz 10 returned almost instantly, to the consternation of snooping observers around the world. The Soviets lied that it had merely been a test rendezvous, not an actual attempt to, you know, get aboard and *use* the space station. There was skepticism at NASA, but, like so much else about the Soviet program, it wasn't until after the fall of the USSR that the facts went public. After shooting into much too high an orbit, Soyuz 10's crew struggled to wrestle the ship back down to Salyut's level. When they finally got there, they couldn't achieve a "hard" dock. So they were brought home—in the middle of the night, rare for a Soviet mission, as though the world's observers wouldn't see their blip on their screens at night. During their sneaky reentry, the Soyuz cabin filled with noxious fumes, and one of the cosmonauts passed out. This wouldn't have happened if they'd been wearing their spacesuits and helmets, but then they'd be too cramped. Let's call that a Soviet compromise. Then, because of the hasty and secretive reentry, they landed off target. This of course was not in itself unusual. They even managed to hit Kazakhstan, which counted as accuracy in the Soviet program. But they also hit a lake. This was a first for the Soviet space program. Frogmen had to be scrambled to rescue them. Unlike all previous cosmonauts who managed to survive their missions, these soggy three got no motorcade and caviar in Moscow. They were quickly shuffled out of the spotlight.

Soyuz 11 took off that June with three new cosmonauts to try again. Originally Alexei Leonov was going to command this mission, but one of his crewmen was diagnosed with possible tuberculosis, so his whole team was replaced. The new team—cosmonauts Georgy Dobrovolsky, Viktor Patsayev, and Vladislav Volkov—had no trouble docking with Salyut. Their troubles started once they were inside. State TV and newspapers regaled the citizens of the Soviet Union with reports of how happy the trio was, grinning as they carried out their "scientific" assignments, clowning around, proud and just so super-glad to be serving the motherland on this historic journey. By the end of the first of their three weeks up there they were household names, new heroes of the people, bringing a bit of the old glory back to the space program.

When the cameras were off, however, it was a very different story. Two people might be able to endure being cooped up in Salyut for a week, even two, if they liked each other very much and had a good supply of sedatives. Three poorly trained middle-aged men—they were forty-three, thirty-eight, and thirty-five—connected only by their government jobs, floating around in there for three weeks, bumping into one another, smelling one another, listening to one another snore, unable to escape from one another's moods and tics and bodily functions, got on one another's nerves. The bickering started early. Volkov, who had flown one previous mission—he was the cosmonaut who once griped that the Cosmodrome represented "the maximal coincidence of inconveniences"—apparently began lording it over the other two, who were first-timers. He tried to take over from Dobrovolsky, who had been appointed mission commander. Other cosmonauts tried to ease tensions from the ground; even Mishin himself stepped in. To make matters worse, the crew refused to carry out the exercise routines they were ordered to do as a way of mitigating the effects of prolonged weightlessness on their muscles and bones. They said that when they used the treadmill in the living/working area it shook the station so badly they were afraid the wings of solar panels that powered everything might fall off. They started to feel weak and slept poorly. Then things got worse than that: sounding highly agitated, they called mission control to report that they smelled smoke. Ground control told them to evacuate to the Soyuz while the situation was assessed. After a while it was deemed safe for them to return to Salyut, but now they were permanently rattled. They argued constantly. Volkov was downright mutinous.

Mission control, to everyone's relief, ordered them to come home early. They buttoned up Salyut and climbed into the Soyuz, wearing only their leisure suits. Which became a problem when they prepared to disengage and a warning light began to blink. Sounding nearly hysterical at this point, Volkov shouted at ground control, "The hatch isn't pressurized! What should we do? What should we do?" Obviously they couldn't disengage if the hatch wasn't completely sealed, unless they were in their spacesuits and helmets. They tried various procedures suggested by the techies on the ground. Nothing worked.

Ground control finally advised them to *tape a piece of paper over the warning light and proceed*. Read that sentence again. Nothing shouts "Soviet space program!" like that single sentence. Their nerves jangling, the crew did what they were told, crazy and stupid as it sounded. They were that desperate to get the hell away from hell. They covered up the light warning them that the hatch was not entirely sealed, disengaged from Salyut, and headed home.

There was no further communication with mission control. The Soyuz 11 DM was tracked reentering the atmosphere as it was programmed to do. It dusted down in Kazakhstan twenty-three days and eighteen hours after lifting off. Rescue teams reached it and opened the hatch. The three men were in there, still strapped to their benches in their leisure suits. They were dead. A quick examination showed nitrogen in their blood, blood in their lungs, and hemorrhaging in their brains—all signs that they had died quick and agonizing deaths when the capsule lost pressure and exposed them to the vacuum of space. They were already dead when the DM, on autopilot, entered the atmosphere. Had they been wearing their spacesuits and helmets, they'd have lived.

After making such a spectacle of the brave, happy heroes of the Soviet Union for three weeks on state TV and in all the state-run newspapers, there was no way the government could cover this one up. Their corpses were flown to Moscow for state funerals. The people of the Soviet Union were in shock. It was the worst space tragedy their government had ever told them about.

The Soviets responded to the calamity with another prolonged period of analyzing what had gone wrong and figuring out fixes. Mishin was blamed for the Soyuz crews' lack of training. No cosmonaut flew for the next two and a half years. It wasn't like they were racing the Americans anymore, so there was no great need to rush into another situation that might end in tragedy and humiliation. When they did go back into space, the practice of sending cosmonauts up there without space gear on had finally ended. Wearing newly designed spacesuits and helmets on takeoffs and reentries meant that only two could comfortably fly in the Soyuz capsule from then on. It was considered a reasonable trade-off—although two key figures, Mishin and, surprisingly, clown car survivor Feoktistov, were against it, arguing that in multi-passenger ships it was better to "ensure collective safety" than protect each individual.

No one ever entered Salyut 1 again. Alexei Leonov called it "haunted." Its orbit was allowed to decay and it burned up entering the atmosphere later that year.

Soyuz 12 in 1973 marked the return of cosmonauts to space. But at the Cosmodrome it's maybe better known for starting another superstitious tradition. The night before they launched, Vasily Lazarev and Oleg Makarov, who had been Leonov's partner on the canceled shot at the Moon, sat in their cabin and enjoyed an oddball Russian movie, *White Sun of the Desert*. It's a kind of spaghetti Western with a distinctly Russian sensibility and sense of humor—an "Eastern," it's been called. Maybe it was all the desert scenery that the cosmonauts, sitting out in the Kyzylkum, liked about it. They took off the next day on a blessedly uneventful flight. Originally they were to dock with Salyut 2, and no doubt do various military reconnaissance chores while aboard it—it was still an Almaz by a different name—but the third stage of the Proton rocket that lofted the space station into orbit blew up, scattering a field of debris right in Salyut 2's path. The impact tore Salyut's solar panels off and penetrated the hull, making it a useless wreck. So Lazarev and Makarov merely flew up and returned. After that, superstitious cosmonauts all wanted to watch *White Sun of the Desert* before lifting off—despite what happened when Lazarev and Makarov flew again two years later. (Cosmonauts also enjoyed a Russian cartoon series, *Nu, Pogodi!* (Well, Just You Wait!), but then apparently everyone in the USSR did. It featured a no-good wolf and a cute bunny whose relationship was like Road Runner and Wile E. Coyote's.)

In April 1975, Lazarev and Makarov took off in what came to be designated Soyuz 18a, headed for two months of military surveillance chores on Salyut 4. The R-7 carrying them took off from the Cosmodrome into a bright blue morning sky, and the ascent was normal for five minutes. Then things went desperately wrong. The core and third stage failed to detach correctly, sending the whole rocket tumbling wildly. A deafening warning siren blared in the capsule, as though the occupants needed to be told they were in trouble. According to Brian Harvey, "Vasily Lazarev reported the problem at once, but ground control would not believe him. For some reason, telemetry did not indicate a fault. 'Abort, abort!' he screamed, 'cut us free!'" He was also reported to shout, "Don't you dumb bastards know what's happened to us?" The abort was finally triggered, the capsule was fired away from the tumbling rocket, then the DM detached from its other two. It "began to fall like a stone" the 192 kilometers to Earth, scoring a fiery trail downward at ballistic speed and at a terrifyingly steep angle, with a force of up to 21 g's crushing the cosmonauts, who said it felt like having a car sitting on your chest. (Fighter pilots experience up to 10 g's max.) They may have blacked out. Fortunately, their parachute deployed automatically. They tried to radio ground control but got no response. There's a legend that ground control could hear them, however, and Glushko went livid listening to them cursing his engines. Looking out their porthole they could see the sun setting on the snowy Altai mountains and worried that they were drifting across Kazakhstan's northern border into unfriendly China. The DM hit a steep snowy mountain-side and began to bounce and spin down it, tossing the cosmonauts around inside, further battering them. Just before they would roll over a cliff their parachute lines got caught in some prickly mountain brush and saved them. They clambered out to frigid darkness. Their first thought was to build a fire in the snow and burn the documents they carried regarding the secret military tasks they would have carried out on Salyut 4. Recovery helicopters spotted them the next day. It turned out they were in friendly Mongolia. There was no safe landing area for the choppers. A rescue team struggling through chest-deep snow to reach them triggered an avalanche. Finally a chopper was able to retrieve them. "They came as close to dying as anyone can and later talk about it," James Oberg concluded. Lazarev had a concussion, broken ribs, and internal bleeding. He never flew again. When he died in 1990, a rumor went around that he'd inadvertently poisoned himself drinking samogon, homemade moonshine.

In 1976, Volynov went back up too, with a Ukrainian cosmonaut named Vitaly Zholobov. They flew in a new and supposedly improved Soyuz 21 and boarded the new, supposedly improved Salyut 5. It was Volynov's second mission in sixteen years as a cosmonaut, Zholobov's first time in space. According to Burgess and Hall, it didn't take long for the two of them to start getting on each other's nerves. Working sixteen-hour shifts in the cramped Salyut space didn't help their moods. Volynov would later complain that he did most of the work, while Zholobov grew weak and lethargic. *Izvestia* would report that they both suffered from "sensory deprivation." There would be other reports that it was nitric acid fumes from a leaking propellant tank that made Zholobov ill. They gritted it out for forty-eight days, which seems truly heroic, and were then permitted to cut the projected sixty-day mission short. By this stage Zholobov was so weak, Burgess and Hall write, that Volynov had to haul him into the Soyuz and strap him down for the return flight.

Which, somehow not surprisingly, did not go well. To start with, the docking clamps would not disengage. One pictures Volynov at this point doing some of the screaming and cursing Komarov was said to have done. He finally managed to get Soyuz 21 free and guide it into a reentry trajectory. Then, "as it descended for a parachute landing near the town of Kokchetav on the Asian steppes of Soviet Kazakhstan, strong, gusting winds caused an uneven firing of the soft-landing rockets," Burgess and Hall write. "As a result the spacecraft impacted hard with the ground, literally bouncing across the ground until it came to a standstill, tipped on its side." It was night. Volynov, weakened after seven weeks of weightlessness, could barely drag himself out of the capsule. Then, as he struggled to pull the limp Zholobov out, Zholobov's helmet "jammed on an obstruction inside the craft," at which point one thinks of Job. The rescue team finally showed up. The Soviet news

agency TASS called the mission "satisfactory," an unusually sub-
dued description for a Soviet spaceflight. Like the Soyuz 10 team
they got no hero's welcome in the Kremlin, which might have been
just as well, since Volynov said it was three days before either of
them could stand up. It was the last mission for them both.

In one respect, Volynov and Zholobov were luckier than the
two who flew in Soyuz 23. Cosmonauts Vyacheslav Zudov and
Valeri Rozhdestvensky flew up to Salyut 5, ran low on fuel trying
to dock with it, failed, and had to make an emergency return in the
dead of night. They landed well off course, in a driving blizzard,
in the middle of the frigid Lake Tengiz in northern Kazakhstan,
the second time cosmonauts splashed down in a lake. Their parachute hit the water, got waterlogged, and sank, dragging the capsule
over on its side, with the exit hatch down, the air vent submerged,
and the radio antennae underwater and useless. The high salinity
of the lake keeps it from freezing over; they were slowly sinking
in icy slush. Even if they could get out it was two kilometers to
shore and they'd freeze to death in the -23 degrees Celsius blizzard.
Nothing to do but hunker and hope for rescue before they froze or
ran out of air, as ice started forming inside the cabin. A recovery
helicopter arrived but took a while to find them in the blinding
snowstorm. Frogmen got to the capsule in rubber boats and struggled to turn it so that the hatch was out of the water. They couldn't.
One of them lost two fingers to frostbite trying. Finally, the next
morning a helicopter dragged the Soyuz through the slush to the
shore. There were fears that the cosmonauts must be dead by then,
but when they popped the hatch they found them unconscious but
clinging to life, suffering from hypothermia and carbon dioxide
poisoning.

★ ★ ★

There were other life-threatening misadventures with Soyuz through the 1970s and into the '80s. But with continuing improvements, it outlived its demonic early reputation and became the workhorse of the Soviet and then Russian space fleets, used by cosmonauts and astronauts from many countries, including the US after the Space Shuttle was mothballed.

In fairness, it should be noted that the Americans' first space station, Skylab, was approximately as awful as Salyut. Launched two years after Salyut 1, it started falling apart during takeoff in 1973, was only in problem-filled operation for twenty-four weeks, then circled the planet for five years an abandoned hulk as its orbit decayed. Its fiery return to Earth in 1979 caused needless worldwide panic. About the size of two school buses, it broke into harmless chunks as it fell. When pieces of it littered the ground around the Australian town of Esperance, Esperance cheekily fined NASA $400 for littering. NASA never paid. But one Esperance teenager earned $10,000 by racing to San Francisco with a Skylab fragment he found and winning a *San Francisco Chronicle* contest to be the first person to bring them a piece of the fallen vehicle. Ranchers around the town were still finding pieces of Skylab on their land in the 1990s.

After Volynov, there wasn't another Russian Jew in space until 2004. NASA's first Jewish astronaut, Judith Resnik, flew on the shuttle *Discovery* in 1984, then died aboard *Challenger* in 1986. More than a dozen Jewish astronauts had flown for NASA by 2021. On October 13 of that year, a ninety-year-old Jewish civilian from Canada joined Jeff Bezos for a brief suborbital flight. His name was William Shatner.

14

HELLO, GOODBYE

In 1973, Alexei Leonov was visiting NASA's space center in Houston. With the race to the Moon over, the American and Soviet space programs were actually prepping for a joint mission. Astronauts and cosmonauts were visiting each other's facilities for the first time, studying each other's equipment, learning each other's languages. The Cold War had never felt so warm.

At a dinner in Houston, Leonov stood to make a toast. He wanted to say, "To a successful future." But because he was still learning English, and maybe because it was not the first toast of the evening, it came out more like "To a sex-full future." In a heartbeat of silence the Americans all looked at one another, then shrugged and said, yeah, we'll drink to that.

When a NASA team made a reciprocal visit to the Cosmodrome, the local secretary of the Communist Party hosted a Kazakh-style banquet that went all Temple of Doom on them. "A boiled, skinned ram's head sat on a platter right in front of me," astronaut Tom Stafford recalled in his memoirs *We Have Capture*. Leonov "picked up the platter and announced that it was 'tradition' that the main host and guest each had to eat one eyeball out of the ram's head. I thought he was joking right up to the moment when he handed me a fork. I'd had a few vodkas by then, so I dug the eyeball out of the head, plopped it in my mouth, and chewed it." One member of the American party turned his head and vomited; two others ran out of the room.

Note Stafford's casual "I'd had a few vodkas by then." It could be the motto for the whole project. If their collaboration with the Soviets taught the astronauts anything, it was respect for vodka.

The idea of the two sides cooperating in space was almost as old as their competing in space. Not long after impulsively announcing a race to the Moon, Kennedy began to have second thoughts, not only when he was told how expensive it would be, but also when it was made clear to him that it was a race not likely to be won until after he'd served his hoped-for eight years in the White House. He'd have to shoulder all the costs just to have some subsequent president—maybe a *Republican* one—get the glory? What was the point of that? For his part, much as Khrushchev enjoyed one-upping the Americans in space, he had even more reason to blanch at the cost of extending the game to the Moon. As early as 1962, the two of them began to have quiet conversations about working together. Speaking to the UN General Assembly on September 20, 1963, Kennedy raised the possibility of "a joint expedition to the Moon." He asked why "should Man's first flight to the Moon be a matter of national competition? Why should the United States and the Soviet Union, in preparing for such expeditions, become involved in the immense duplications of research, construction, and expenditure?" Khrushchev did not immediately respond—because, he later admitted, cooperating with the Americans would reveal how threadbare the Soviet space program really was. Congress and the media reacted poorly, and Kennedy began to backtrack. Two months later, he was killed in Dallas. In the speech he was going to give that day he would have reverted to the need for America to dominate and lead in space, no feel-good stuff about cooperation.

The idea was revived in 1972, during the Cold War thaw the Nixon administration called détente and Brezhnev's called *razryadka*—Khrushchev's "peaceful coexistence" come back around. The two leaders met in Moscow that May, mostly to talk nuclear arms reductions, but also a collaboration in space. Nothing so grand as a joint trip to the Moon—Apollo's last mission there, Apollo 17, would be in December, and the Soviets had long since dropped it as a goal. This would be just a "handshake in space"—a rendezvous of an Apollo vehicle and a Soyuz. Under the prosaic banner ASTP (Apollo-Soyuz Test Project), rocketeers on the two sides began prepping. The project was simple in theory. Two cosmonauts would go up in a Soyuz, three astronauts in an Apollo capsule. They'd rendezvous, dock, and spend a couple of days being friendly with one another on live TV in the name of détente/razryadka. Then they'd separate and come back home.

In practice, of course, it was more complicated, and took three years of planning, rehearsal, and diplomacy. Two veterans were chosen to lead the cosmonaut and astronaut teams: Leonov, who'd survived the worst the Soviet space program could throw at him, and Stafford, who had flown in both Gemini and Apollo capsules. They had actually met before, in not the most auspicious way, at the

Moscow state funeral for the three Soyuz 11 cosmonauts in 1971, to which Stafford had been sent to represent American sympathies, a small but notable early exercise in détente. In his memoirs Stafford remembers Brezhnev "weeping openly." At the airport the next morning his plane was delayed while a Russian general insisted he share a caviar-and-vodka breakfast, with the requisite many toasts. David Scott, despite the trouble he got into with Apollo 15, was assigned to the American team in a support role. The interpreters for the Soviet team were KGB. The American team included four officers who'd been trained in spycraft for the abandoned MOL program.

Space diplomacy turned out to be complicated. Neither side could make territorial claims on space—it was the ultimate neutral zone—so much thought went into how the cosmonauts and astronauts would greet each other when they docked. Neither side could appear to be "welcoming" the other into "their" space. They were coached and choreographed to keep it all mutual and equal, like travelers simply encountering each other at an inn on the road to the stars. Years later, Leonov would remember the phoniness of the "highly scripted show" of good fellowship they put on in orbit while "literally thousands of weapons were pointing at each other" not so far below. An anthropologist writing about it some thirty years later wondered at the "almost weirdly positive" virtue signaling of the whole affair.

The encounter had to be conducted so that it didn't remind the world that the perpetually touchy Soviets had lost the space race. And it couldn't conflict with the Soviets' continuing bluff that their space technology was on par with the Americans'. It was far from it, as everyone knows now, but it was in fact this project that showed the Americans for the first time how truly ramshackle and inferior the Soviets' equipment was. The grip-'n'-grin had to happen in a lower orbit than the astronauts were used to, because the Soviets couldn't put the Soyuz any higher. Also, the Soviets were still mostly controlling their vehicles from the ground. It worried them that the American cowboys insisted on piloting their Apollo craft themselves. The Soviets were aware of how fragile their machinery was. During the long period of prep, after failing to convince the Americans that the docking should be done by ground control, they reminded Stafford over and over to dock *gently*. The situation was so ripe with opportunities for bawdy jokes that the participants were schooled in not using language suggesting that the Soyuz would adopt a "passive," "female" role and Apollo the male, "active" one.

As they trained, the two teams made several visits to each other's countries. The Soviets were delighted that the gormless Americans were so willing to let them see their space facilities in Houston and the Cape, but they were reflexively horrified by the idea of returning the favor. It violated their mania for secrecy—plus they worried, as Khrushchev had, that it would reveal how hokeypokey their gear was. But the Americans, and especially Stafford, insisted: "I never fly in a spacecraft that I haven't been in on the ground." So, very reluctantly, the Soviets opened up Star City and the Cosmodrome. The cosmonauts, who had been told all their lives that America was not the land of plenty it pretended to be, couldn't get over what a land of plenty it was. The cars, the clothes, the shops, the movie theaters, the homes, and the food—food everywhere, all kinds of food, so much food that the Americans were often careless and wasteful with it. Leonov asked Stafford why, since Americans had it so good, all the news on their TVs was so bad. Stafford had no answer. They loved Disneyland and Disney World. They kept joking that nothing in America was real. They'd see ducks in a pond and claim that they were animatronic. Maybe they'd read the French semiotician Jean Baudrillard, who famously wrote that "Disneyland is presented as imaginary in order to make us believe that the rest is real"—which is to say, you entered the *real* Disneyland when you drove out of the parking lot. (The Matrix movies were inspired by his ideas and some of Philip K. Dick's.)

At the Cosmodrome, Leninsk had grown and changed a lot since the early 1960s. There was a new hotel with tennis courts and a pool, a movie theater, new shops, and buses—*air-conditioned* buses. At Star City the Soviets built a new three-story hotel especially to welcome the Americans, but called it Hotel Cosmonaut anyway. The rooms were big but already dowdy, and Stafford's was full of flies. The Americans were sure their rooms were bugged, as all their accommodations surely were. So Stafford stood in the middle of his room and loudly proclaimed to no one, "The Russians are very wonderful and hospitable people, but it's too bad they decided to be so cheap about this hotel. There's not even a fly swatter." Coming back a few hours later, they found that "every room had a fly swatter. Better yet, every fly in my room had been killed and dumped in the unflushed toilet." After that they started "talking to the walls" whenever they wanted something. They complained to the walls about the lousy Russian beer, and superior Czech, East German, and Egyptian beers appeared that afternoon. "'Talking to the walls' worked better than room service."

Because of all the problems and disasters with the Soyuz and Salyuts—the ones they knew about, anyway—NASA and the astronauts wanted assurances that the Soviets were actually capable of pulling this off. Their visits to Star City and Baikonur weren't terribly confidence-building. Until now they'd had no clear idea of how low-budget and slapped-together the Soviet space program was. Now they saw firsthand. *Newsweek* reported that NASA techies were "dismayed" and "distressed to learn that the Soviets have skimped" on training and equipment. Their simulator for the crucial docking tunnel through which the astronauts and cosmonauts would pass into each other's vehicles, for instance, was a simple ten-foot cylinder, like an XXL toilet paper tube, useless for training. "The Soviets think they can fly this mission by the seat of their pants," one of the NASA techies complained. "That's light years away from our approach, which is to plan for all possible contingencies."

Right Stuff, meet Wrong Stuff. Yippee-ki-yay, comrades.

The Soviets didn't have a Disneyland to show their visitors. They took them to Tashkent and to Kaluga, Tsiolkovsky's hometown. "We had been picked up from the hotel early that morning and told we would have breakfast on the way," David Scott recalled. "I don't know what I expected, a small restaurant, I suppose. But after an hour or so the bus pulled over to the side of the road. We were led off the bus and into a small clearing in the forest. 'We'll be having breakfast now,' one of the Russians announced. With that he started smoothing out a newspaper on the ground on which was laid raw sturgeon, a bottle of vodka and paper cups. It was a far cry from scrambled eggs and bacon, tough on the stomach. We'd been up quite late the night before, but we braced ourselves and started drinking vodka from the leaking cups."

Leonov recorded another time the Soviets poured too much liquor into Scott, when the astronaut led a US delegation visiting Star City. "As always, our wives took great trouble to prepare a good meal for our guests. We moved from apartment to apartment, introducing

David and his companion to other members of the cosmonaut corps, and drank many toasts to the success of the Apollo–Soyuz mission and to further cooperation between the United States and the Soviet Union. It was clear David thought it would be impolite not to accept any aspect of our hospitality, but as the number of glasses of vodka and whiskey we drank mounted he seemed to grow a little unsteady on his feet. The wife of one of our cosmonauts comforted him by saying that everyone feels a little unwell from time to time. But it was clear it was time to have our guests escorted back to their hotel."

In the midst of all this, there was a sea change at OKB-1 in 1974: Vasily Mishin was fired. He had never been popular as Korolev's successor, and never very good at it. If the Kremlin had been paying the sort of attention to the program that Khrushchev used to, there's not a doubt Mishin would have been axed long before 1974, maybe as early as Apollo 8's trip around the Moon, certainly no later than when Armstrong set foot there. With dreadful irony, the man who took over the design bureau that Korolev had founded and led to glory was none other than his old nemesis, Glushko. Adding insult, Glushko ended OKB-1, folding it into his own design bureau under a new name, Energia. One of his first acts was to officially end any more work on Korolev's disastrously flawed N1. The Chief Designer's ashes must have been spinning in their Kremlin Wall niche.

After all that, the actual ASTP mission, in July 1975, was a bit anticlimactic. The Soviet team, following what was by then an established tradition, watched *White Sun of the Desert* the night before lifting off. The next morning they all pissed on the right rear wheel of the bus ferrying them to the launchpad. In orbit, as Stafford jockeyed the Apollo capsule toward the waiting Soyuz, Leonov reminded him to be gentle. On docking, Stafford played the country ballad "Hello, Darlin'" by fellow Oklahoman Conway Twitty. It was a subtle hint that the Soyuz was the darlin' female partner in this union, as well as his response to a joke Leonov had made about understanding English well enough, but not Stafford's "Oklahomski." They played each other more songs during the mission, the most poignant of which might have been War's "Why Can't We Be Friends?" For their first meal together, Leonov offered the Americans a toast of cosmonaut food tubes labeled *Vodka*. The Americans were excited. Three years of dealing with the Russkies had given them quite a taste for it. But when they squeezed the tubes into their mouths it turned out just to be borscht. Leonov laughed at their disappointed reactions. He'd pranked them.

When they weren't fooling around, the two teams played their choreographed parts for the several TV cameras, which beamed the footage down to earthbound audiences who weren't exactly rapt. Space was humdrum by 1975. They cooperated on some small scientific experiments, which were of no interest to anyone but the scientists who designed them.

At the end of the mission the astronauts got Leonov back for his prank. When the Apollo undocked and was maneuvering away, Leonov radioed to say farewell. Instead of answering, the astronauts played a tape of water splashing and young ladies giggling. It sounded like the Americans were having a hot tub party. Leonov knew, of course, that as superior and luxurious as the Apollo was compared to Soyuz, it didn't include party facilities. But for a moment he was thrown.

As it turned out, the handshake in space was more a goodbye than a hello. Even as Leonov and Stafford were pranking each other in orbit, negotiations to reduce nuclear weapons were stumbling.

Jimmy Carter entered the White House in 1977 focusing on the Soviets' human rights abuses—quite rightly, since the whole of the Soviet Union could be characterized as one big human rights violation. The Nixon and Ford administrations had been willing to look the other way in pursuit of détente. Relations cooled, then went into the deep freeze in 1980, the year of Brezhnev's invasion of Afghanistan and the election of the staunchly anti-Soviet Ronald Reagan. There would be no more handshaking in space until after the Soviet Union fell. Maybe the most lasting effect was the new brand of Apollo-Soyuz commemorative cigarettes.

The ASTP Apollo could have been called Apollo 18 but wasn't. The Apollo program had officially ended in 1972. This one was Apollo leftovers, a generic Apollo, and the very last one used. As happy as President Nixon had been to bask in Apollo 11's glory—after all, it had been his idea before Kennedy's—by '72 he was reducing NASA's budget and its ambitions. No more guff about lunar colonies or going on to Mars or any of those other dreams that the rocketeers, both American and Soviet, had cherished since they were starry-eyed kids devouring sci-fi. Nixon, like Brezhnev, pointed them at projects closer to home, more affordable, with more obvious military uses. Hence the Space Shuttle, which he authorized that year. "It will revolutionize transportation into near space, by routinizing it," he said.

It certainly did the latter. Except for its two terrible disasters, the thirty years of the Shuttle's workaday operations, from April 1981 to July 2011, killed any lasting sense of adventure in space exploration.

Space might still be the final frontier, but its Wild West years were over. No more giant leaps, just a trudge to and from the job. Astronauts might as well have been swinging lunch pails and wearing hard hats. Shuttles did put up a number of classified military intelligence packages, but the public knew nothing about that at the time, and much remains secret. Otherwise it earned its keep as a Hubble repair truck and the Space Shuttle bus, running passengers to and from the space stations Mir and ISS.

Which raises what may be the most interesting point about the Shuttle: the way the Soviets reacted to it. In 1974 their leading military men, angling for funding as always, had a closed-door meeting with Brezhnev and wound the aging dipsomaniac up with dire lies about the "space bomber" the Americans were building. They'll be able to steal our satellites! They could attack our space stations! They'll be able to fly over Moscow and drop nukes on us! We must counter with a space bomber of our own! Brezhnev was alarmed and envious enough to approve it, despite what looked like would be an enormous bill.

Designers, including Glushko, developed new space bomber proposals, but they were all rejected in favor of the time-honored Russian approach: They'd simply buy or steal the Americans' blueprints for the Shuttle and copy them in their own Counter-Shuttle. NASA made that laughably easy with the feckless decision to keep its Shuttle documentation unclassified and openly available. The Soviets educated themselves on Shuttle design from documents they requested straight from the US Government Publishing Office. They also siphoned thousands of pages of Shuttle documents from the infant internet, where they'd been uploaded from government, university, and contractors' computers. The Soviets saved themselves years of research and billions of rubles on the Buran (Blizzard) as they called it.

The Americans finally woke up in the early 1980s to how much technical data the Soviets were mining from them. The tip came from a disgruntled KGB agent named Vladimir Vetrov, to whom his Western contacts gave the code name FAREWELL—rather ominous, considering that he would eventually be caught and executed. Rather than stop making their data public, which would alert the Soviets that the jig was up, the Americans carried out a bit of clever spycraft: They inserted false and misleading data into the stream. For instance, it's believed the hoodwinked Soviet engineers put defective tiles on the hull of the Buran, which could trigger serious problems during reentry.

Regardless of the shortcuts they took, the Soviets were still hobbled by the usual supply chain problems and general disorganization. The development of the Buran crept along through the 1980s. The Buran would not quite be a note-for-note copy of the Shuttle. The main difference was that like everything else in the Soviet space program it was hugely heavy, more a Space Anchor than a Space Shuttle. The Soviets were still lacking the lightweight "space-age" alloys available to NASA. Hence the need for a giant new rocket, Glushko's Energia, just to get the Buran off the ground. But there was another transportation problem: how to get the Buran from Glushko's shop outside of Moscow to Baikonur in the wilderness of the Kazakhstan steppe. It was too big and heavy to go by rail or road. The answer was yet another piece of gargantuan engineering, the Antonov An-225, a jet that would fly the Buran piggyback to the Cosmodrome. NASA's Shuttle was also transported on the back of a jet, a modified 747, but it was smaller in most every way, because the Shuttle wasn't an

Anchor. The An-225 was built at the Antonov shop in Kyiv and given a Ukrainian name, Mriya, Ukrainian for Dream. It was a behemoth, the biggest aircraft ever built, with six engines, thirty-two wheels of landing gear, and wings wider than the Statue of Liberty is tall. It's been noted that the cargo hold was so long (142 feet) that the Wright brothers could have made their first flight (120 feet) entirely inside it.

Meanwhile, in 1983 American President Ronald Reagan significantly ramped up Soviet stress levels when he announced the Strategic Defense Initiative (SDI), a research program to develop a high-tech umbrella against Soviet ICBMs. Over the next several years American defense contractors hoovered up billions of public dollars working on glamorous ideas for space lasers, a hypervelocity rail gun that blew itself off its rails every time it was fired, and other exotic weaponry. Reagan's critics denounced SDI with the nickname Star Wars, though his interest in lasers may have begun with another movie, a 1940 B picture called *Murder in the Air*. He starred in that one as the Secret Service agent Brass Bancroft who was on the trail of saboteurs with a new terror weapon, the "inertia projector," a beam weapon not unlike a laser. It's online.

The Soviet military interpreted SDI not as a defensive shield but as a new first strike initiative. Or at least that's how they presented it to Kremlin leaders, who agreed to shell out billions of rubles on, not surprisingly, trying to copy the new American gizmos. It would be too much to say that the Star Wars race bankrupted the Soviet Union and pushed it into its grave, but it certainly contributed. The Soviet Union was in no condition to engage in a new space-race-plus-arms-race with the Yanks. The Afghan war dragged on. Buran and Energia were already soaking up billions. Profits from the vast Siberian oil fields that had been propping up the economy sank. After Brezhnev's death in 1982, the leadership was unstable; his doddering successor Yuri Andropov only lived to 1984, and *his* successor, Konstantin Chernenko, only made it into 1985. Mikhail Gorbachev took over that year with the quixotic mission of making the Soviet Union more open, democratic, and capitalist while retaining its Marxist-Leninist core. Not surprisingly, he mostly just generated more confusion and dysfunction than ever. In trying to reform the USSR, he was killing it.

The Buran was finally ready, sort of, in 1988. The An-225 carried it to Baikonur, where the mighty Energia booster lifted it into orbit that November. After more than a decade in development Buran was still unfinished, including life support, so it was unmanned and flown by remote from the ground. It made two orbits and came in for a nice landing back at the Cosmodrome.

And never flew again. By 1988 Gorbachev's Soviet Union was broke and broken. Among many other budget cuts, funding for Glushko's bureau dried up. That one flight of Buran was Glushko's last hurrah. He died the following January, age eighty-one. The Buran was rolled into a hangar at the Cosmodrome, and assembly was halted on two others that were in mid-production, even though one of them, which was officially the Burya (Tempest) but nicknamed Ptichka—Birdy—was said to be as close as 97 percent finished.

On July 27, 1991, cosmonauts Anatoly Artsebarski and Sergei Krikalev went EVA on the space station Mir (Peace). The first module of

Mir had been launched in 1986 and it grew in segments like a tinker toy over five years. To celebrate Mir's completion, Artsebarski and Krikalev attached the USSR's red hammer-and-sickle flag to it.

Less than a month later, the end of the hammer-and-sickle era began. On August 19, conservatives in the government and military who were fed up with Gorbachev's reforms attempted a coup, arresting him and rolling tanks into Moscow. It failed in three days, but two Soviet republics, Ukraine and Kazakhstan, took advantage of the confusion to declare their independence. Among other things, this meant that the Cosmodrome was now on foreign soil. Through that autumn Krikalev watched from 220 miles up as the Soviet Union teetered toward the brink. Artsebarski flew home and was replaced by Alexandr Volkov, who was a good partner for Krikalev. The two of them were among the most experienced Soviet cosmonauts ever and had previously served on Mir together. Both reveled in long missions. On his first flight in 1988 Krikalev had spent five months on Mir. He fell in love with his future wife, a flight technician on the ground, while speaking with her by radio. He had gone back up to Mir in May for a planned eight-month mission. Krikalev actually liked being on Mir, which set him apart from most others who experienced it. Mir was much bigger than the old Salyuts, about the size of six school buses hooked together, and yet somehow just as cramped, and stinky and noisy.

Krikalev and Volkov were up there when Gorbachev announced the end of the Soviet Union on December 25, 1991, leaving the two citizens hanging, literally. They were suddenly men without a country, holding useless passports and Communist Party membership cards. Krikalev, who had already been up there for eight months, was not sure when or even if he'd be able to return to his wife and infant child. Merry Christmas, Comrade Krikalev. For a few months they were afterthoughts floating in space while the intense politics of a Communist empire that had disassembled itself into eighteen sovereign republics were worked out. This included negotiations between Kazakhstan and what was now the Russian Federation over use of the Cosmodrome. Volkov and Krikalev finally returned to Earth in March, landing in Kazakhstan. As he was pulled out of a Soyuz capsule after 312 days of weightlessness, Krikalev looked limp as a marionette whose strings had been cut, which was metaphorically apt. He'd gone up to Mir a patriotic Soviet citizen and Communist Party member, and returned to a world where none of that was valid anymore.

The Russian Federation, working its deals with Kazakhstan, continued to keep Mir in operation—barely. The transition from Marxist empire to a supposed free market democracy came with tremendous political, social, and economic chaos. By 1993 the Russian space agency, which would come to be called Roscosmos, was forced to collaborate with NASA and other foreign entities on sending non-Russian astronauts and scientists up to Mir, raking in very much-needed foreign currency for resuming the handshaking in space. One of the unexpected consequences of the Soviets having copied the Shuttle back in the 1980s was that Mir was perfectly ready to handle visits from the Shuttle. It had a docking port designed for the Buran, easily adapted.

Like Aeroflot in the 1950s and '60s, Mir alarmed foreign users. Astronaut Jerry Linenger wrote about his harrowing time aboard the "dark, ramshackle, incredibly cluttered" vehicle in his book *Off the Planet*, with the telling subtitle *Surviving Five Perilous Months Aboard the Space Station Mir*. When he went to Star City for training in 1995 he was dismayed by how decrepit the facilities were. Buildings were crumbling, public bathrooms were repulsive, most spaces were barely lit or heated. Classroom training was little more than ancient Gagarin-era engineers standing in front of blackboards droning on and on in Russian. While he and other foreign visitors worked out in the bare-bones gym, the Russian cosmonauts hid in a steam room with the requisite bottle of vodka. The era of glamorous celebrity cosmonauts was over. They were just working stiffs now, grateful to have jobs in the roiling post-Soviet economy. Onboard Mir their relationship to ground control was as class-conscious and adversarial as the one between Ripley and the Parker-Brett duo on the *Nostromo*, and just like on the *Nostromo*, their bosses used alternating bonus offers and pay-cut threats to prod them into grudging compliance.

Mir had been designed for five years of operation. When Linenger got there in 1997 it was eleven years old and showing many signs of age and neglect. Crew members, who sometimes lived onboard for more than a year, spent most of their time fighting a losing battle of housekeeping and repairs, like sailors in a leaky open boat constantly bailing for their lives. The clutter was astounding. Cables and conduits that should have been routed through walls or overhead snaked along the floor and slithered through the hatches. Which meant that in case of emergency—a fire, a module venting atmosphere—you couldn't easily retreat into another module and seal the hatch behind you. The emergency would follow you. Electrical systems regularly failed; lights flickered, and the comms system was so buzzy with static that crew members could barely hear one another. More humble but important equipment like the low-gravity toilets failed. A deafening alarm like an old-fashioned klaxon periodically blared *a-OO-ga a-OO-ga* and the crew scrambled to figure out what was wrong now. Fans and other gear kept up a rattling and banging background noise that got more nerve-racking the longer you were there, and the whole contraption groaned and creaked worryingly every time it passed from the sunny side of the Earth to the night side. It had accumulated eleven years of discarded junk, and one module ostensibly dedicated to astrophysics had become a filthy garbage dump through which the weightless crew had to propel themselves, gagging at the stench. The rest of the place smelled moldy and dank, like a locker room. That was because of the basketball-size globules of bacteria-loving water that accumulated in the walls from having people aboard for eleven years breathing and sweating in the sealed, low-gravity environment. The place was an orbital Petri dish.

Mir used canisters filled with a chemical mix that, when activated, generated breathable oxygen. Unfortunately, the chemicals also had a bad tendency to burst into flames. Linenger had barely settled in when the *a-OO-ga* alarm started shouting and just such a fire filled the space station with acrid smoke. As the crew swam through the smoke, reaching blindly for extinguishers and oxygen masks, the flame burned so intensely it started melting a hole in the hull. Had it burned through, the air would have been sucked out immediately and the six crew members would likely have died. It burned itself out, which was lucky, because emptying their extinguishers on it seemed to have no effect. The Russian authorities downplayed it as a small fire easily contained, and blandly lied to the crew about how safe the canisters were. NASA played along.

Michael Foale, the British-American astronaut who replaced Linenger on Mir, almost died in a different way. Except for occasional visits by a Shuttle, Mir was supplied by smaller, remote-controlled cargo ships, Soyuzes that had been scooped out to make room for bringing supplies up and hauling junk out. The remote control system was made in the now independent republic of Ukraine, which gleefully, not to say vindictively, charged the Russians out the wazoo for it. The Russians decided to save money and have a cosmonaut on Mir manually bring the Soyuz in instead. It would be a fantastically complex and stressful operation even if all the equipment involved worked perfectly, which it never did. Inevitably this improvised procedure failed. A Soyuz supply ship flying ten times faster than it should crashed into Mir, breaching the hull of one module so that the air started rushing out. As Foale hurried to seal the hatch to that module he had to chop through the clutter of hoses and cables that jammed it, sparks flying, causing a dangerous power outage. He barely got the hatch closed in time. Mir had been knocked into a crazy spin that took some time to stop. The Russian authorities, naturally, put all the blame on the cosmonaut.

Mir was finally put out of everyone's misery in 2001. It had been replaced by the International Space Station, which received its first crew in 2000—one of whom was none other than Sergei Krikalev. The question now was how to deorbit the abandoned Mir safely. Since Sputnik flew in 1957 Earth has become increasingly shrouded by an ever-denser fog of man-made objects. At the start of 2022, nearly five thousand satellites were up there, and more were being launched all the time. But they represent less than 10 percent of the cloud. The rest is millions of pieces of junk, from as small as nuts and bolts, flecks of paint, and bits of copper wiring to whole dead satellites—Vanguard 1 from 1958 was still up there in 2022—and the upper stages of old rockets.

The tiny objects or "micro-debris" pose a unique problem, because they can't be tracked. Today's space agencies use "debris avoidance maneuvers" to try to keep spacecraft, space stations, and satellites out of harm's way while they test methods of clearing up at least some of the junk.

Larger objects like Mir can be instructed to reenter the atmosphere, and told where to impact. Looking for the remotest place on the planet, engineers mapped the "oceanic pole of inaccessibility"—the area on the planet farthest from any land—in the middle of the South Pacific, which they called, in a nod to Jules Verne, Point Nemo. It's a watery wasteland, surrounded by about fourteen million square miles of featureless ocean. The nearest land, the Pitcairn Islands, is more than fourteen hundred miles away. Which meant that if you were standing on the deck of a ship at Point Nemo, the nearest humans were nowhere on the planet, but on whatever space station happened to pass overhead, just a couple hundred miles above you. But you were unlikely to be on a ship at Point Nemo anyway, since it's far from any shipping lanes. No one goes there. That's the point of Point Nemo.

There's no better place on Earth to bring space rubble down in a planned descent and let it crash into the sea, where it will sink to the bottom—which at Point Nemo is about thirteen hundred feet down. NASA, the Soviets and Russians, the Japanese, and other space agencies have been using Nemo as an underwater space graveyard since the early 1970s. Mir is down there. The Chinese space station Tiangong-1, which failed after five years in orbit, was guided to a watery grave at Point Nemo in 2016. At least half a dozen Salyuts are there, and SpaceX began to use it for discarded rockets. Space agencies say there's little worry that all this junk will pollute the Pacific, claiming that fuel and other toxic substances get burned up during reentry.

Really? Tell that to the people who live downrange from Baikonur and Plesetsk. They're still busy spaceports, where hundreds of UDMH-fueled Protons were launched by the Soviet and then Russian space programs after 1965. The first and second stages dropped downrange of the cosmodromes, where the people who eked out a threadbare subsistence in poor hamlets supplemented their incomes by salvaging and recycling this space junk. They stripped gold and titanium from the wreckage to sell on the black market, and turned sections of crashed rocket stages into boats called *rocketa*. In handling these materials they exposed themselves to the residual Devil's venom. According to studies in the 1990s, nine out of ten people in certain downrange areas around both cosmodromes displayed heightened rates of pathologies ranging from liver damage to rickets, and overall death rates were as much as 30 percent higher than average. UDMH also turned a vast area of the Kazakh steppe into an ecological wasteland.

After 1991 the Russian Federation was forced to lease the Baikonur Cosmodrome from Kazakhstan. There wasn't much play in the crashed Russian economy, and the facilities at the Cosmodrome and the town of Leninsk—officially renamed Baikonur in 1995, even though the original Baikonur was still a few hundred kilometers away—crumbled. Reminders of the glory days were hard to find. The one Buran that flew and the two that didn't stood neglected in hangars, gathering rust and grime through the 1990s. In 2002 the roof of the Buran's hangar caved in, wrecking it and killing eight workers. In 2021 some graffiti artists sneaked into the big, gloomy shed where Ptichka and the other incomplete one were abandoned, caked in years of dust, looking like big junked cars. They tagged the hulks with cranky slogans like *Before flying to the stars, we must learn how to live on Earth.* That kicked up a wave of Buran nostalgia in Russia. They were freshly painted, the shed swept and dusted, and security guards were posted.

Here the story takes a funny turn. In 2011, a Kazakh businessman named Dauren Musa had taken private ownership of the Cosmodrome. The Russians paid the rent to him from then on. After the graffiti incident in 2021, Roscosmos offered to buy Ptichka. Musa said he'd only part with it in exchange for another rare and valuable artifact: the head of the last Kazakh khan, Kenesary Kasymov. Kasymov was beheaded in 1847 while fighting for Kazakhstan's independence from the tsar. His head was sent to Russia, probably St. Petersburg. Roscosmos was willing to negotiate, but there was a hitch: no one in Russia could remember where Kasymov's head was. Evidently it was an item of far more value to Kazakhs than to Russians.

Vladimir Putin, the former KGB operative, never hid how nostalgic he was for the Soviet Union. He called its end "the greatest geopolitical catastrophe" of the twentieth century. His tyrannical autocracy might be described as twenty-first-century Stalinism without the Communism. Russia's invasions of the former Soviet lands of Georgia, Crimea, and the rest of Ukraine conveyed a perverse sort of sentimentality.

When the Red Army entered Ukraine in February 2022, one of the first targets was as symbolic as it was strategic. After the Buran program was terminated, the gigantic An-225 jet that had carried it to the Cosmodrome sat in its hangar at Kyiv's Hostomel airfield—well, mostly "in," since its vast tail couldn't fit inside—through most of the 1990s. Then it saw some service as an Antonov Airlines cargo plane, the largest in the world, the pride of Ukraine, culminating with carrying huge loads of Covid-19 supplies during the pandemic.

That February, as one of the first acts of the invasion, missiles fired by Russian forces smashed Mriya, the Ukrainian Dream, where it sat at the airfield. How ironic that Vladimir Putin, the latest in a centuries-long line of Russian despots, pining for the lost Soviet empire, saw fit to destroy one of its few positive legacies. Plumes of black smoke drifted erratically skyward, signaling that whatever else had changed, the Wrong Stuff remained.

ACKNOWLEDGMENTS

Thanks to my friend Bill Monahan, who got the point of this book from day zero, and gave me the perfect title.

Thanks to Diane Ramo, who was an enormous help the whole time I worked on it.

Thanks to my friends Laura Lindgren, Ken Swezey, and Richard Byrne, who all had some great ideas.

Thanks to Ben Adams, Melissa Veronesi, Pete Garceau, Susie Pitzen, Bart Dawson, and everyone else at PublicAffairs. And to Anthony Mattero and everyone at CAA. Nice working with all of you.

FURTHER READING

Brzezinski, Matthew. *Red Moon Rising: Sputnik and the Hidden Rivalries That Ignited the Space Age.*

Burgess, Colin, and Chris Dubbs. *Animals in Space: From Research Rockets to the Space Shuttle.*

Burgess, Colin, and Rex Hall. *The First Soviet Cosmonaut Team: Their Lives, Legacy, and Historical Impact.*

Burrows, William E. *This New Ocean: The Story of the First Space Age.*

Carlson, Peter. *K Blows Top: A Cold War Comic Interlude, Starring Nikita Khrushchev, America's Most Unlikely Tourist.*

Cherrix, Amy. *In the Shadow of the Moon: America, Russia, and the Hidden History of the Space Race.*

Chertok, Boris. *Rockets and People*, Vols. 1–4.

Coleman, Fred. *The Decline and Fall of the Soviet Empire: Forty Years That Shook the World, from Stalin to Yeltsin.*

Daniloff, Nicholas. *The Kremlin and the Cosmos.*

D'Antonio, Michael. *A Ball, a Dog, and a Monkey: 1957—The Space Race Begins.*

Day, Dwayne A., John M. Logsdon, and Brian Latell, eds. *Eye in the Sky: The Story of the Corona Spy Satellites.*

Doran, Jamie, and Piers Bizony. *Starman: The Truth Behind the Legend of Yuri Gagarin.*

French, Francis, and Colin Burgess. *Into That Silent Sea: Trailblazers of the Space Era, 1961–1965.*

Gagarin, Yuri. *Road to the Stars.*

Golovanov, Yaroslav. *Our Gagarin: The First Cosmonaut and His Native Land.*

Hall, Rex, and David J. Shayler. *The Rocket Men: Vostok and Voskhod, the First Soviet Manned Spaceflights.*

Harford, James. *Korolev: How One Man Masterminded the Soviet Drive to Beat America to the Moon.*

Harvey, Brian. *The New Russian Space Programme: From Competition to Collaboration.*

Harvey, Brian. *Soviet and Russian Lunar Exploration.*

Jenks, Andrew L. *The Cosmonaut Who Couldn't Stop Smiling: The Life and Legend of Yuri Gagarin.*

Khrushchev, Nikita. *Khrushchev Remembers.*

Khrushchev, Nikita, ed. by Sergei Khrushchev. *Memoirs of Nikita Khrushchev*, Vols. 1–3.

FURTHER READING

Khrushchev, Sergei. *Nikita Khrushchev and the Creation of a Superpower.*

Linenger, Jerry M. *Off the Planet: Surviving Five Perilous Months Aboard the Space Station Mir.*

MacDonald, Fraser. *Escape from Earth: A Secret History of the Space Rocket.*

McDougall, Walter A. *The Heavens and the Earth: A Political History of the Space Age.*

Metcalfe, Robyn Shotwell. *The New Wizard War: How the Soviets Steal U.S. High Technology—and How We Give It Away.*

Mieczkowski, Yanek. *Eisenhower's Sputnik Moment: The Race for Space and World Prestige.*

Oberg, James E. *Red Star in Orbit.*

Oberg, James E. *Uncovering Soviet Disasters: Exploring the Limits of* Glasnost.

Phelan, Dominic. *Soviet Space Secrets: Hidden Stories from the Space Race.*

Rhea, John, ed. *Roads to Space: An Oral History of the Soviet Space Program.* Translated by Peter Berlin.

Riabchikov, Evgeny. *Russians in Space.*

Schrad, Mark Lawrence. *Vodka Politics: Alcohol, Autocracy, and the Secret History of the Russian State.*

Scott, David, and Alexei Leonov. *Two Sides of the Moon: Our Story of the Cold War Space Race.*

Shayler, David J. *Disasters and Accidents in Manned Spaceflight.*

Siddiqi, Asif A. *Challenge to Apollo: The Soviet Union and the Space Race, 1945–1974,* Vol. 1.

Siddiqi, Asif A. *Soyuz 1: The Death of Vladimir Komarov.*

Stafford, Thomas P., with Michael Cassutt. *We Have Capture: Tom Stafford and the Space Race.*

Stone, Tanya Lee. *Almost Astronauts: 13 Women Who Dared to Dream.*

Suvorov, Vladimir. *The First Manned Spaceflight: Russia's Quest for Space.*

Taubman, William. *Khrushchev: The Man and His Era.*

Tereshkova, Valentina. *In Her Own Words: The First Lady of Space.*

Titov, Gherman, and Martin Caidin. *I Am Eagle!*

Vladimirov, Leonid. *The Russian Space Bluff: The Inside Story of the Soviet Drive to the Moon.*

Volkogonov, Dmitri. *Autopsy for an Empire: The Seven Leaders Who Built the Soviet Regime.*

Walker, Stephen. *Beyond: The Astonishing Story of the First Human to Leave Our Planet and Journey into Space.*

Waterlow, Jonathan. *It's Only a Joke, Comrade!: Humour, Trust and Everyday Life Under Stalin.*

Wolfe, Tom. *The Right Stuff.*

John Strausbaugh is a well-known author of history books. His titles include *Victory City*, *City of Sedition*, and *The Village*. A former editor of *New York Press*, he has written about history and culture for the *New York Times*, the *Washington Post*, *Evergreen Review*, the *Wilson Quarterly*, and other publications.

PublicAffairs is a publishing house founded in 1997. It is a tribute to the standards, values, and flair of three persons who have served as mentors to countless reporters, writers, editors, and book people of all kinds, including me.

I. F. STONE, proprietor of *I. F. Stone's Weekly*, combined a commitment to the First Amendment with entrepreneurial zeal and reporting skill and became one of the great independent journalists in American history. At the age of eighty, Izzy published *The Trial of Socrates*, which was a national bestseller. He wrote the book after he taught himself ancient Greek.

BENJAMIN C. BRADLEE was for nearly thirty years the charismatic editorial leader of *The Washington Post*. It was Ben who gave the *Post* the range and courage to pursue such historic issues as Watergate. He supported his reporters with a tenacity that made them fearless and it is no accident that so many became authors of influential, best-selling books.

ROBERT L. BERNSTEIN, the chief executive of Random House for more than a quarter century, guided one of the nation's premier publishing houses. Bob was personally responsible for many books of political dissent and argument that challenged tyranny around the globe. He is also the founder and longtime chair of Human Rights Watch, one of the most respected human rights organizations in the world.

· · ·

For fifty years, the banner of Public Affairs Press was carried by its owner Morris B. Schnapper, who published Gandhi, Nasser, Toynbee, Truman, and about 1,500 other authors. In 1983, Schnapper was described by *The Washington Post* as "a redoubtable gadfly." His legacy will endure in the books to come.

Peter Osnos, *Founder*